The Book On
The Shape of What's Coming

Why Everything Feels Broken,
And What's Actually Taking Shape
Beneath the Surface

The Book On Series
Alex Nova

Published by The Book On Publishing, 2025.
First edition. June 29, 2025.

.

I0458875

Website: https://thebookon.ca
Substack: https://thebookonpublishing.substack.com/

While every precaution has been taken in the preparation of this book, the publisher assumes no responsibility for errors or omissions, or damages resulting from the use of the information contained herein.

THE SHAPE OF WHAT'S COMING

First edition. June 29, 2025.

Written by Alex Nova.

The Book On Series

The Book On Life Unscripted

The Book On Risk Management in Payments

The Book On AI for Everyday People

The Book On Relationships

The Book On Master The Algorithm

The Book On Saying No

The Book On Community Led Strategy

The Book On The Myth of Multitasking

The Book On The Burnout Blueprint

The Book On The Digital Reboot

The Book On The Shape of What's Coming

The Book On Strategic Obsession

The Book On High-Stakes Thinking

The Book On Artificial Leverage

The Book On Clarity

The Book On Uncertainty

The Book On Operational Excellence

The Book On Escape

Table of Contents

READ THIS FIRST 6

DEDICATION 8

EPITAPH 9

PREFACE 10

SECTION I: THE QUANTUM THRESHOLD 12

CHAPTER 1: BEFORE THE STORM BREAKS 12

CHAPTER 2: REALITY AS CODE 18

CHAPTER 3: WHY SHAPE IS STRONGER THAN SUBSTANCE 24

CHAPTER 4: MASTERING UNCERTAINTY 29

CHAPTER 5: SECRETS OF EDGE-STATE PHYSICS 34

CHAPTER 6: THE RETURN OF THE SUPERMATERIAL 39

CHAPTER 7: BUILDING SYSTEMS THAT DON'T FAIL 44

CHAPTER 8: WHO IS BETTING ON THE FUTURE 49

CHAPTER 9: CROSSING A TECHNOLOGICAL THRESHOLD 54

CHAPTER 10: WAKING UP TO A NEW REALITY 59

SECTION II: TOPOLOGICAL MACHINES 66

CHAPTER 11: WHEN THE MACHINE CHANGED 66

CHAPTER 12: BRAIDING LOGIC INTO MATTER......................... 70

CHAPTER 13: WOVEN MATH: THE NEW COMPUTATION 75

CHAPTER 14: GEOMETRY WITHOUT FRAGILITY 79

CHAPTER 15: TECHNOLOGY BUILT TO LAST 84

CHAPTER 16: MATTER THAT SPEAKS LOGIC........................... 88

CHAPTER 17: MEMORY WITHOUT STORAGE 92

CHAPTER 18: THE SHAPE OF MODERN INTELLIGENCE 96

CHAPTER 19: ENGINES OF EMERGENCE 100

CHAPTER 20: THE NEW ARCHITECT OF INTELLIGENCE 104

SECTION III: INTELLIGENCE WITHOUT ILLUSION 109

CHAPTER 21: THE DEATH OF COMMAND.............................. 111

CHAPTER 22: SENSING WITHOUT EYES 116

CHAPTER 23: THE INNER MODELS WE BUILD 120

CHAPTER 24: LEARNING FOR SURVIVAL 125

CHAPTER 25: HOW WE LEARN WITHOUT LABELS 131

CHAPTER 26: THINKING WITH THE BODY 137

CHAPTER 27: PREDICTION IS THE PURPOSE 142

CHAPTER 28: WHY EXPLAINABILITY IS A MYTH 148

CHAPTER 29: WHEN MACHINES NEGOTIATE 153

CHAPTER 30: SEEING OURSELVES IN THE SYSTEM 159

SECTION IV: THE QUANTUM CITIZEN 166

CHAPTER 31: EXPLORING THE EDGE: WHAT THEY WON'T TELL YOU 166

CHAPTER 32: THE TRANSLATOR'S EDGE 171

CHAPTER 33: QUANTUM FOR EVERYONE 176

CHAPTER 34: LEVERAGING THE FRINGE 182

CHAPTER 35: BUILD IT TO UNDERSTAND IT 187

CHAPTER 36: HOW INFLUENCE HIDES IN PLAIN SIGHT 192

CHAPTER 37: CREATING QUANTUM SPACE 197

CHAPTER 38: DIGITAL CITIZENSHIP IN THE QUANTUM ERA 203

CHAPTER 39: YOUR FIRST 100 MOVES IN A QUANTUM WORLD 208

CHAPTER 40: YOU ARE THE SYSTEM 213

SECTION IV: ENTANGLED FUTURES 220

CHAPTER 41: MORALITY IN A QUANTUM AGE 220

CHAPTER 42: ENTANGLED MINDS 225

CHAPTER 43: DEMOCRACY VS. SINGULARITY 231

CHAPTER 44: THE LOGIC OF THE GODS 236

CHAPTER 45: THE UNIVERSE AS COMPUTATION 241

CHAPTER 46: END OF SCARCITY 246

CHAPTER 47: WHAT COMES AFTER CAPITALISM 251

CHAPTER 48: THE GREAT UPLIFT OR THE FINAL DIVIDE 255

CHAPTER 49: WHAT LEGACY WILL WE LEAVE? 260

CHAPTER 50: THE BECOMING 264

EPILOGUE 267

GLOSSARY OF KEY CONCEPTS 270

REFERENCES & FURTHER READING 275

ABOUT THE AUTHOR 276

ABOUT THE PUBLISHER 277

ACKNOWLEDGMENT OF AI ASSISTANCE 278

Read This First

This is not a book designed to entertain you. It's not here to charm, to soothe, or to hold your hand. It won't dazzle you with stories, metaphors, or motivational fluff. What you're having is a tool, an instruction manual written for people who are serious about learning, executing, and thinking at a higher level.

Every book in The Book On Series is built on a single premise: clarity beats complexity. We believe that when you strip away the noise, the emotions, the marketing spin, and the cultural rituals of "self-help," what's left is raw, unembellished instruction. That's what these books offer.

They are dry by design. Not because we don't care about language or narrative, but because when you're building something that matters, you don't need more distractions. You need a clear architecture. Mental scaffolding. Direction that respects your intelligence.

Each title in this series takes on a specific domain: decision-making, clarity, strategy, leverage, and uncertainty, and drills deep, not in sweeping generalizations, but in applied frameworks. These are books for builders, operators, founders, tacticians, and thinkers—people who don't just consume knowledge but operationalize it.

You'll find no chapter-long anecdotes here. No self-congratulatory memoirs. No bullet-point platitudes. Instead, what you'll get is structured insight: argument, example, application. The tone is direct. The prose is sober. The ideas are designed to be lifted out and used.

You won't be coddled, but you won't be misled either.

There's a place in the world for lyrical, emotional, story-driven books, and this isn't that place. This is a workspace. A blueprint. A conversation for people who are ready to act, not just absorb.

We respect your time and your intellect.

Welcome to The Book On Series.

Dedication

For the readers who never waited for permission, the builders without blueprints, the thinkers who questioned the frame, and the citizens who dared to feel complexity instead of fleeing it.

This is for you.

The world you sensed is real. The future you feared is open. The story, at last, is yours.

- A.N.

Epitaph

For the Age Between Worlds

Here lie the last certainties of the classical world, the straight lines, the fixed truths, the comforts of control.

They served us once, but not anymore.

What rises in their place is not chaos, but pattern. Not collapse, but complexity.

We walked blind into the quantum dawn and opened our eyes not to machines, but to mirrors.

May we remember: the future was never a machine to fear; it was a story waiting for authors. And we were always the ink.

- A.N.

Preface

We're not living through ordinary times.

Every so often, history takes a turn, one that doesn't just change our tools but reshapes how we understand ourselves, how we relate to each other, and how we imagine the future. These moments go deeper than technology; they're about civilization itself. Today, with the rise of quantum science, topological thinking, and networked intelligence, we're standing at such a threshold. Yet, if you only followed the headlines, you might miss it entirely.

What you'd likely see instead is a world spinning faster than it can explain itself: systems growing more complex, narratives fracturing, futures feeling prewritten and out of reach, built by a small elite, interpreted by algorithms, and disconnected from most of us.

The Shape of What's Coming was born out of a belief that this dominant story is not only incomplete, but also dangerous.

This book is meant as an invitation, not from a technologist to other specialists, but from a translator to fellow citizens. It aims to make the strange more familiar, to help you reclaim a sense of agency in a time when even meaning feels uncertain. Together, we'll explore emerging ideas and realities, quantum computing, moral topology, civic entanglement, decentralized systems, and the more profound philosophical questions they provoke when the world no longer behaves the way we once assumed it would.

But this isn't just a book about what's changing. It's about how we live through that change, with curiosity, clarity, and care.

You don't need a PhD to follow this journey. You don't need wealth or credentials to take part. What's required is a willingness to think in new ways, to feel what's at stake, and to act with intention.

There won't be simple answers here. But what you will find is a map, a more honest one, that reflects the actual contours of the future, not the comfortable fictions of the past.

Because in a quantum world, nothing stands alone. Mind, matter, memory, meaning, they're all connected. Once we begin to see that clearly, the real question isn't what the future holds. Which future will we choose to create, together?

Thank you for opening this book.

Let's begin.

Section I: The Quantum Threshold

Chapter 1: Before the Storm Breaks

Somewhere, a teenager scrolls past a viral dance and a new particle discovery in the same breath. A bus lurches forward. A server hums in a climate-controlled vault. The world seems normal, ordinary.

There's a strange quiet before every great technological upheaval. In the moment, everything feels ordinary. Markets rise and fall. Streets stay busy. Students graduate into careers. The future still looks like the present, just with smaller phones and better cameras. But beneath that surface, something foundational has already shifted, and soon, it will change the rules for everyone.

You are living in that quiet now.

We tend to imagine revolutions as explosions, loud, visible, and immediate. But in truth, the biggest ones begin in silence. A theory that rewrites our understanding of nature. A material behaving in a way no one predicted. A way of storing or transmitting information that doesn't just improve what came before, it renders it obsolete.

This book is about one of those moments. And you are reading it on the very edge of the wave.

Not far from where you are, physicists are bending electrons in strange materials to make them move only along the edges of atoms. Engineers are crafting circuits where information isn't encoded in silicon chips, but in the *geometry* of the material itself. Computers are being built not with logic gates, but with quantum states that exist in two or more places at once.

This isn't science fiction or speculation; it's already happening, just not yet on the front page of your favourite news app.

What's emerging now is more than a new kind of device. It's a new class of reality-shaping technology. One that doesn't just compute faster. It redefines *what is computable*. One that doesn't just carry electricity. It determines *how* electricity flows or doesn't. One that doesn't just use materials. It *designs materials* that behave according to the invisible principles of quantum logic and mathematical topology.

This isn't a better version of the world we know. It's the opening act of a world we don't yet recognize.

What Topology and Quantum Physics Are About

To appreciate what's happening, you don't need a PhD in physics. You need a shift in perspective.

You need to understand that nature doesn't care about our intuitions. Reality is not built out of solid chunks. It's built out of relationships. It's not organized like a machine. It's organized more like a computation, or perhaps a thought.

Quantum mechanics, the most successful and strange theory in all of science, tells us that particles don't have fixed positions or velocities. They exist in fields of probability until we observe them. They can be entangled across vast distances, such that changing one immediately changes the other. They are not this *or* that; they are this and that until forced to choose.

And topology, a branch of mathematics that studies how things hold together when stretched, bent, or twisted, shows us that some truths don't depend on size or shape. A donut and a coffee mug, each with one hole, are the same in the eyes of topology. What matters is not the form, but the structure of relationships.

When these two ideas, quantum and topological, meet, we begin to see a world where stability is built into shape, where logic lives in space itself, and where information is protected not by shielding it, but by *entangling it into the structure of the universe*.

For years, one thing has held back quantum computing: fragility. Qubits, the basic units of quantum computation, are delicate. They lose coherence quickly. A tiny fluctuation in temperature, a stray electromagnetic pulse, or even a curious measurement can destroy the state and erase the information. That's why current quantum computers, while promising, remain mostly laboratory curiosities.

The answer to this fragility isn't more brute force. It's a better design. And this is where topological materials enter, like a quiet revolution.

Imagine storing a secret not in one location, but across a *knot* in space, a structure so robust that it can't be untied without destroying the whole thing. Imagine encoding information in the *twist* of a particle's path, such that noise and error can't find a foothold. This is what topological quantum computing promises: a system where quantum information is no longer fragile, because it's no longer *just local*. It's structural.

This is not a metaphor. It's already being prototyped in labs, by Microsoft, by small startups, by physics departments on the bleeding edge. They are working not just to build better qubits, but to *change the fabric of computation itself.*

We've seen this pattern before. In the 19th century, electricity was a parlour trick, amusing, mysterious, but useless. Then came the lightbulb. Then the power grid. Then the world.

In the early 20th century, the discovery of the electron and the structure of the atom seemed like abstract, even absurd theories. But within a generation, those theories gave birth to transistors, lasers, and the digital revolution.

Today, we're watching something similar unfold. But it's harder to see, because the shift is not from one machine to a better machine. It's from *machines* to *materials*, from *circuits* to *shapes*, from hardware to topological reality.

Most people won't notice until the wave has already broken.

What This Book Is (and Isn't)

This book is not a textbook. It's not a collection of speculative tech jargon. It is a guide, a roadmap for intelligent, curious people who sense that the world is about to change in ways more profound than anyone is saying.

You don't need to understand the mathematics. But you will need to shift your mindset. You'll need to get comfortable with paradox, with probability, with uncertainty, and with the radical idea that reality itself is computable and may soon be rewritten.

Over the coming chapters, you'll learn:

➢ How topological materials are designed, and why their properties are not optional

➢ Why quantum logic is not just strange, but valuable, and why the world we live in may already depend on it

➢ What topological conductors do, and why they might make future electronics near-perfect

➢ Who is investing in this future, and what do they know that the public doesn't

➢ And most importantly, what you can do with this knowledge, whether you are a student, a creator, a policymaker, or just a citizen of the 21st century

In every era, some people see it coming. They're not always the loudest, or the best funded, or even the most celebrated. But they're watching closely. They're reading between the lines.

They're noticing when something invisible begins to hum beneath the floorboards of the present.

That's where you are now.

You are standing at the edge of something immense and almost entirely unrecognized.

This is not the end of the digital age. It is its evolution. And what's coming next will not only change how we compute. It will change how we understand time, space, connection, and even **truth**.

The wave is forming. You are in the calm just before it crests, and this is your invitation to ride it.

Chapter 2: Reality as Code

When you touch your phone's screen, a million things happen beneath the glass: light pulses. Electrons move. Signals bounce between microprocessors. None of it feels like magic anymore. That's the quiet triumph of the digital world; it has become invisible. So seamless that we no longer notice it working.

But behind this seamlessness is an idea that took centuries to take hold: the world can be understood, and built, as code.

That idea is now about to mutate. We are transitioning from a world programmed by humans to one where nature itself behaves like a computational system. Where matter isn't just shaped, it's instructed, where reality, at its core, might be less like a stage… and more like a simulation running on rules far stranger than we imagined.

This chapter is about what happens when that realization crosses from physics labs into public consciousness. And what it means when the building blocks of existence start to behave like software.

The ancient world believed in elements: earth, fire, water, and air. Later, science replaced those with atoms and molecules, then electrons and quarks. Each era moved the "stuff" of the universe into smaller, more abstract layers. But what we often forget is that all of these layers, at their deepest level, are not solid at all. They are rules.

Matter isn't stable. It is a process. What you experience as a thing, a rock, a hand, a thought, is a pattern of behaviors, obeying physical laws. Electrons orbit not because they're little planets, but because that is what the mathematics of quantum fields demands. Particles don't bounce off each other like billiard balls. They interfere, entangle, and collapse because they are following instructions.

It took physics centuries to accept this. But now, at the bleeding edge of research, a radical consensus is emerging: the universe might not just be describable as a computation. It might be computation. In the language of some researchers, nature isn't like a computer; it is a computer. The laws of physics are its operating system. Particles are its data structures. And time? That may be the ticking of an invisible processor.

If this sounds abstract, consider this: much of modern theoretical physics is now written not in terms of matter, but in terms of information. The fundamental limits of what can happen in the universe, including speed, quantity, and accuracy, are being described in terms of bits, entropy, and logic gates.

Black holes, once thought to be mysterious dead zones, are now believed to behave like massive information processors, erasing, scrambling, and possibly even storing the data of the matter that falls into them. The entire universe's expansion is being modeled with code-like precision. And at the most minor scales, quantum systems behave exactly like probabilistic programs, ones where observing a variable executes part of the logic.

This is why quantum computers don't just seem magical. They make natural sense in a world where reality behaves like software. They don't simulate the universe. They mirror it.

To understand what's coming next, we need to think less like classical scientists and more like programmers of a cosmic architecture.

In the last chapter, we glimpsed how topology gives materials unusual properties, like the ability to conduct only along the edge, or to preserve a quantum state despite noise. But what makes that truly extraordinary is this: those behaviours are not just chemical. They are logical.

You can program them.

Scientists are now designing materials where the geometry itself encodes information, where an electron's spin is locked to its direction of motion, where flipping a state in one location flips its partner miles away. These aren't gimmicks. They are the logical equivalents of IF statements and memory registers, but made of matter.

This is not just about computing faster. It's about the merging of software and hardware at a level never seen before. Imagine if your chair could sense the temperature and restructure its surface for comfort. Or your jacket could adjust its stiffness to protect your bones during a fall. Imagine a bridge that could respond to stress by rerouting molecular bonds to compensate, not through machinery, but through instructions built into the material itself.

That's where we're heading. A world where we don't just write software. We print it into reality.

Now imagine you grow up in a world like that. A world where your tools aren't just responsive, they're alive with logic. Where everything from clothing to walls can think, remember, and adapt. You would start to see the world less as a collection of objects and more as a sea of instruction sets.

This shift in perception is not optional. It's already happening. Young engineers are learning to think in quantum states. Material scientists are learning to believe in topological spaces. Investors are learning to think in probabilistic outcomes, not certainties. And perhaps most profoundly, ordinary people are beginning to sense, even if they don't yet understand, that the rules are changing.

Because when the world becomes code, everything becomes programmable. Not just devices. Not just systems. But experiences, identities, even memories.

What, then, is real?

There is a temptation, especially in the West, to think that to understand something is to control it. But as we peer deeper into this code-shaped reality, we're finding something else: complexity that resists domination. Quantum systems can't be fully observed without being changed. Topological states can't be altered locally; they require transformation at the system level.

This is not a universe of clean levers and switches. It is a universe of entangled processes, where participation changes outcomes.

To operate in that world, we must let go of the idea of mastery and instead learn collaboration. We will build systems not by dictating every instruction, but by guiding their evolution, sculpting their parameters, letting the logic of the material express itself.

We won't program the future. We'll co-compose it.

As in every previous revolution, from writing to printing to computing, there is a new kind of literacy forming. It's not the ability to memorize equations or recite facts. It's the ability to see the world as systems, to understand patterns across layers of reality, to manipulate abstraction as easily as matter.

This is the literacy of the next century. Those who learn to think in terms of code, structure, and probability will shape the tools, policies, and philosophies of what comes next. Those who don't may feel increasingly out of sync with a world that no longer behaves as expected.

This is not a judgment. It's an invitation.

When people ask what's next for technology, they expect answers like faster chips, more intelligent AI, and better batteries. But the deeper answer is stranger: what's next is a world made of logic, embedded in space, woven into time, alive in the flow of electrons and energy.

What's next is a civilization that doesn't just use technology, it becomes technological in its architecture, its assumptions, its relationships. And in that world, knowing how to build, think, and live in systems of code, whether digital, biological, or quantum, will be as important as knowing how to speak.

The world as we knew it was made of things. The world that's coming is made of instructions. Learn to read them. And you learn to rewrite the future.

Chapter 3: Why Shape is Stronger Than Substance

Imagine taking a rubber band and twisting it into a knot. You stretch it, squeeze it, even warp it into something unrecognizable. But the knot stays. You can contort the whole object beyond recognition, yet the loop, the structure, the *idea* of the knot, it holds. What you're witnessing is not strength in the traditional sense. It's a different kind of resilience. Not toughness, but topology.

Topology is a branch of mathematics that doesn't care about measurements. It doesn't care about size, distance, or angle. What it cares about, and what it preserves, are relationships. Connections. Continuity. To a topologist, a coffee cup and a donut are the same thing. Each has one hole. Each can be smoothly deformed into the other without tearing or gluing anything. This isn't a visual joke. It's a principle that runs beneath our most advanced technologies, and soon, our most powerful machines.

The deeper scientists look into the rules of the universe, the more it seems that shape matters more than substance. And that realization is quietly redefining how we build everything.

For centuries, we've assumed that what matters most is what something is made of. Steel is stronger than wood. Diamonds are more complex than glass. Silicon is faster than copper. But that worldview, while useful, is beginning to crack. Increasingly, the most resilient systems we're building, from quantum computers to fault-tolerant circuits, don't just rely on the stuff they're made from. They rely on how that stuff is *arranged.*

Topology gives us a way to encode logic, memory, and even identity into the fabric of a system. Not by adding more parts. Not by shielding it from error. But by building it in such a way that its essential features can't be broken, not by noise, not by damage, not even by measurement.

A topological feature is like a melody played across a room. You can move the furniture, change the temperature, even paint the walls, but if the notes are still played in the same sequence, the song survives. The shape of the information matters more than the components carrying it.

In the physical world, this has profound consequences.

In the 1980s, researchers observed something strange in a class of materials subjected to extreme cold and powerful magnetic fields. Electrons moved not randomly, as expected, but in perfect, discrete steps, as if obeying some hidden code. The effect was so precise that it became a new way to define electrical resistance. It was called the Quantum Hall Effect, and at first, it was a curiosity.

But then came the realization: this behaviour wasn't chemical. It wasn't magnetic. It was topological.

The electrons were confined to the edges of the material, gliding along the boundaries with near-perfect immunity to defects. The interior was insulating, a dead zone. But the edge? It was alive. Conductive. Protected. Like a road that can't be blocked, no matter how many potholes you try to dig. It was as if the material had developed its internal logic. A built-in rule that said: this current *will flow*, and you cannot stop it.

This marked the dawn of what we now call topological phases of matter. Not a state like solid, liquid, or gas, but a new kind of behavior governed by geometry and quantum mechanics. It was a breakthrough, and a warning. Because it meant the rules we thought governed materials were incomplete.

There was a deeper layer. And it had always been there.

The implications were breathtaking. If you could build a material where certain behaviors were protected not by force, but by structure, then you could build machines that didn't fail under stress. Systems that retained their properties regardless of what you threw at them. Memory that didn't need to be refreshed. Circuits that didn't short under noise. Logic that could endure beyond the imperfections of the real world.

This is where topology meets computation. And where physics began whispering the possibility of a new kind of computer, one where information isn't stored in fragile bits or volatile transistors, but in topological states that can't be disturbed without tearing apart the whole system.

In other words, a computer that doesn't forget. Even when shaken.

Topology doesn't offer us magical materials. It provides us with a new language for thinking about design. A way to build structures whose functionality is guaranteed not by what they're made of, but by how their parts are connected.

In biology, we see echoes of this. DNA strands twist and loop in ways that protect them from degradation. Neural networks form pathways that reinforce themselves with use. Even ecosystems, the oldest information systems on Earth, evolve webs of redundancy and feedback that survive extinction events. These are topological patterns, not engineered designs. Nature found them first.

Now we are catching up. Scientists are building synthetic materials whose conductivity depends on their shape. Circuits are being woven like fabric, not printed like boards. Experiments are underway to create memory that lives in braids, literal twists of quantum fields, rather than in silicon.

It's as if we've discovered a new class of architecture: structures that think, because their form enforces behavior.

All of this may seem abstract. But its consequences are immediate. Our world is built on fragile systems, from financial algorithms to climate models, from power grids to communication networks. These systems, for all their sophistication, are often brittle. They break under pressure. They fail in the face of the unexpected.

Topology offers an answer. Not a way to prevent chaos, but a way to build through it. A way to construct systems that don't rely on perfect inputs, that survive noise, distortion, even partial destruction, because their essential truth is encoded in how they hold together, not in the precision of any one part.

This is the principle behind topological quantum computing. But it's also the principle behind everything from resilient cities to emotional intelligence. The idea that some things don't break because they're hard, but because they are shaped to survive.

We are entering a future shaped by this principle. And the people who understand it early will be the ones who know how to build not just more intelligent machines, but more imaginative worlds.

It's easy to think of knowledge as facts and systems as functions. But what topology teaches us is that structure is destiny. That is how things connect, and it determines how they behave. And that the most resilient truths are often invisible until you stretch them.

Somewhere, right now, in a lab or a startup or a government think tank, someone is designing a system whose behaviour will be protected not by force, but by form. Someone is encoding logic into braids, memory into loops, resilience into geometry. And someone else, maybe you, is just beginning to see what that means.

We are only beginning to understand the implications of this. But already, one truth stands clear: The future will not just be faster or wiser. It will be topological.

Chapter 4: Mastering Uncertainty

There's a story physicists like to tell. It goes like this: you put a cat in a box, along with a device that has a fifty percent chance of releasing poison within the hour. The mechanism is triggered by the decay of a single radioactive atom, a purely quantum event. You close the lid. You wait.

According to the equations that describe the situation, the atom is now both decayed and not decayed. The device has both released the poison and not. And so, until you open the box, the cat is both dead and alive.

It's absurd. It's unsettling. And it's the closest we've come to describing how the universe works.

Quantum mechanics is not a theory of particles and waves. It is a theory of possibilities. A framework that says: what exists is not just what *is*, but what *could be*, entangled in a shimmering field of probability until observation collapses the choices into one outcome.

This isn't science fiction. It's science. And the world you live in depends on it more than you think.

If classical physics was the age of certainty, of apples falling, pendulums swinging, and levers moving, quantum physics is the age of doubt. But not confusion. Not randomness. Rather, a kind of structured uncertainty. A precision in ambiguity. A logic of not-yet.

It began, strangely enough, with light. In some experiments, light behaves like a wave, spreading out and interfering with itself. In others, it behaves like a particle, arriving in discrete packets called photons. Which is it? Both, somehow. And neither, until measured. The same is true of electrons. Of atoms. Of whole molecules.

This duality, the fact that something can act as two things at once, defies everyday logic. But in quantum logic, it is not only accepted. It is necessary.

An electron doesn't orbit a nucleus the way a planet orbits a star. It exists in a cloud of possibilities. A quantum computer doesn't store a bit as 0 or 1, but as a superposition of both, a fluid, changing mixture of states that only becomes definite when you ask it to. And sometimes, just by asking, you change the answer.

This is the world as it is, beneath appearances. A world not built of certainties, but of potentials, not of fixed facts, but of evolving probabilities.

You don't need to believe this. You're already living inside it. Every GPS signal that guides your car relies on quantum corrections to account for relativity. Every smartphone chip runs on principles derived from quantum tunneling. Every laser, every MRI, every LED, they all depend on this "illogical" physics being correct. And it is. Predictively. Repeatedly. Uncannily.

But the real magic begins when we stop using quantum principles to explain the world and start using them to build it.

Quantum computing is the most famous example, and also the most misunderstood. A quantum computer is not just a faster calculator. It is a fundamentally different machine. Where a classical computer must try each possible solution one by one, a quantum computer can explore many simultaneously. Not by doing more at once, but by reshaping the space in which solutions reside.

Imagine solving a maze not by trying every path, but by folding the maze so that the entrance and exit align. That's the kind of logic we're dealing with. Not brute force, but subtle geometry. Not linear progression, but interference, cancellation, and entanglement.

It sounds mystical. It's not. It's math. But it's a kind of math that behaves like poetry.

There's a paradox in all of this. Quantum systems are robust precisely because they are fragile. A quantum state can hold multiple truths, but the act of looking forces a choice. Measurement becomes a kind of violence, an interruption. You can read the data or preserve it. Not both.

This is why building quantum computers is so hard. Any interaction with the outside world, such as heat, light, or sound, can collapse the delicate state. Coherence vanishes. The computation is lost. Engineers are racing to find ways around this, using refrigeration, shielding, and error correction algorithms more complex than the systems they protect.

But what if we didn't need to protect the system by force?

What if we could build a system that preserved quantum information not by isolating it, but by embedding it in a structure?

This is where topology returns. And why quantum + topological = revolution.

The union of quantum and topological principles offers a path to stability. If you can encode information not in a fragile local state, but in the shape of an entire system, a loop, a braid, a knot in the quantum field, then that information becomes hard to destroy. You'd have to alter the whole structure to erase it. And that's much harder than flipping a bit.

This is the dream of topological quantum computing. Not faster machines. Not just weirder machines. But more reliable ones. Machines that can perform quantum logic without falling apart under the weight of their elegance.

We are still in the early days. But the logic is sound. The prototypes exist. The promise is staggering.

It's not just about solving math problems. It's about simulating molecules, cracking unbreakable codes, and designing materials that nature itself has never built. It's about thinking in a new medium, one where uncertainty is not a flaw, but a feature.

There is, perhaps, something philosophical at play here. For centuries, we've built our world on the assumption that reality is knowable, predictable, and fixed. If you had enough data, you could forecast everything: weather, behavior, markets, and futures.

Quantum theory laughs at this. Not cruelly. Not mockingly. But gently, as if to say: you were never meant to know everything. You were meant to navigate uncertainty.

To think quantum is to accept that truth is often plural. That opposites can coexist. That observation changes the observed. That knowledge comes with a cost. And that the best we can do is act with humility inside systems too complex to see in full.

This isn't defeat. It's an invitation. To grow wiser, not just smarter. To make peace with the not-yet. To build technologies that reflect the complexity and fragility of the world they inhabit.

Right now, the future of quantum technology, of topological computation, is in superposition. It is both real and not-yet. Both imminent and uncertain. The breakthroughs are here. The scale is not. The physics is ready. The world is not.

But this is how every revolution begins.

And if you've read this far, you are already ahead of most because you're learning the grammar of the next age, a logic not of certainty, but of unfolding.

The rest of this book will show you how that logic is becoming matter. How quantum rules are being embedded in the materials we use. How computation is escaping its digital box and entering the physical world.

We will move from wavefunctions to wires. From theory to design. But never forget: the foundation of it all is uncertainty. And it is beautiful.

Chapter 5: Secrets of Edge-State Physics

It would be easy to think that all innovation begins with power, more energy, more speed, more force. But in the world of topological conductors, something stranger is proper. Here, stability doesn't come from strength. It comes from shape. And perfection doesn't arise from order. It emerges from the edges.

This is the story of how we learned to bend the laws of matter not by fighting nature, but by understanding a subtle truth hidden in its fabric: that in certain materials, under certain conditions, electrons will move in ways that seem almost enchanted. They will glide effortlessly along the outer rim of a structure, refusing to scatter, refusing to stop, even if the interior is full of chaos.

The path of least resistance, it turns out, is not always through the center. Sometimes it's around the edges. And that is where the future of electronics is quietly forming.

In most materials, electricity flows like water through a pipe. It moves because there's a push, a voltage, and it encounters resistance because the pipe isn't perfect. When electrons bump into atoms, they scatter, heat the system, and energy is lost. This is why wires get hot, why batteries run out, and why we've always had to trade efficiency for scale.

But in the mid-1980s, researchers discovered something shocking. In fragile materials, chilled to near absolute zero and exposed to magnetic fields, electrons stopped acting like particles. They began behaving like a single, unified wave. More than that, the current stopped flowing through the middle and started hugging the outer edge, moving in tight, well-defined paths. It no longer mattered if the material was full of impurities or irregularities. The current didn't care. It went around them. This was not a minor breakthrough. It was a complete redefinition of what conduction could mean.

For the first time, scientists had witnessed edge states, paths where electrons are bound not by chemistry or force, but by topology. And in these edge states, something miraculous happened: the electrons could travel without resistance. No collisions. No loss. A perfect glide, immune to noise.

This phenomenon came to be called the Quantum Hall Effect, and it marked the beginning of a new field. A quiet revolution in the design of materials. And the birth of topological conductors.

Most of our lives are organized around the idea that boundaries are limits. They divide, they restrict, they separate the inside from the outside. But in topological systems, boundaries do something different. They become carriers of truth.

The edge of a topological insulator, a special kind of quantum material, is where all the action is. The interior may be inert, insulating, or useless for current. But the outer rim is alive. It conducts electricity in a way that's both stable and astonishingly precise. No matter what you do to the inside, the edge keeps working. No matter how dirty or damaged the system becomes, the edge stays clean.

It's as if the system has chosen to protect its most valuable behavior by outsourcing it to the border.

This isn't an accident. It's a consequence of deep mathematical rules. When the quantum states of a system are shaped in specific ways, when they "twist" through space and time, they create protected modes at the boundary. These modes cannot be erased without altering the entire system's topology. They are global features, immune to local noise.

And in that, they embody a powerful idea: that the most resilient truths are not hidden in the heart of things. They are traced along the edges.

The implications are immense. Imagine a wire that doesn't lose energy. A circuit that doesn't heat up. A processor that doesn't decay with time. These are not fantasy goals. They are engineering targets already being pursued in labs and startups around the world.

Topological conductors promise a world where heat loss is no longer the price of speed, where the fragility of electrons is no longer a limiting factor, and where the rules that typically degrade performance, noise, friction, and scattering are irrelevant.

Because the electrons never enter the battlefield. They glide above it, along the boundary, in perfect synchrony.

Already, prototypes exist. Materials like bismuth selenide or engineered graphene layers show signs of these effects. They are being studied, tuned, layered, and stacked. And while the room-temperature version of these materials is still in development, the logic is sound. Nature permits this. The equations demand it. And that means, eventually, the market will catch up.

What begins in a vacuum chamber today could shape the power grid of tomorrow.

This is a turning point in how we think about material science. For decades, we've engineered from the inside out, building better atoms, purer compounds, cleaner lattices. But topology teaches us to design from the outside in. To shift the focus from perfection to structure. To realize that the most important thing about a system might be how the edges are arranged.

It's a design principle that extends beyond physics. In software, edge conditions determine whether programs crash or survive. In ecosystems, the boundaries between environments are where biodiversity flourishes. In cities, cultural innovation often begins at the margins.

Now, the same is becoming true in electronics. The center can be imperfect, so long as the boundary is right.

This challenges everything we thought we knew about performance, stability, and design. It invites a new kind of imagination. One where function emerges from form. Where protection comes from structure. And where the edge is not the end, but the source.

Topological conductors are not yet widespread. You don't yet carry them in your pocket. But you will. Because they solve a problem we've been stuck with for over a century: the trade-off between performance and resilience.

The moment we no longer need to sacrifice energy to make systems faster, or precision to make them scalable, everything changes. Data centers shrink. Supercomputers run cooler. Batteries last longer. And maybe most importantly, quantum computers become *practical* because the same principles that guide edge conduction also stabilize quantum information.

We are watching the birth of a new class of materials. Those that not only allow current to flow but also *guarantee* it. Ones that hold not just electrons, but principles, encoded not in their ingredients, but in their geometry.

If this sounds abstract, it won't for long. The market is moving. The labs are glowing. The patents are being filed. A generation of engineers is being trained not just to build faster machines, but to shape matter in a way that makes failure impossible.

Not by fighting disorder. But by rendering it irrelevant.

And when that happens, when the world learns to compute, to communicate, and to power itself by the rules of topology, we will look back and realize it all began at the edge.

Chapter 6: The Return of the Supermaterial

There was a time when bronze reshaped the world. It allowed empires to rise, weapons to evolve, and cities to be built from something more enduring than wood and bone. Then came iron. Then steel. Each new material brought not just strength, but structure, a new way to organize human effort. Then came silicon, and with it, the digital age.

But something unusual is happening now. The materials that will define this century won't be forged in heat. They won't be mined from the earth. They will be engineered, atom by atom, designed to obey principles far stranger and far more potent than anything in our industrial past.

We are entering the era of supermaterials. And unlike their predecessors, these are not just stronger or lighter. They are logical. Quantum laws and topological constraints shape them. They are designed not just to survive pressure, but to process information, manipulate energy, and stabilize behavior at the level of reality itself.

This is not material science as we've known it. This is a new kind of matter.

Traditionally, a material was valuable for what it could *endure*. How much weight could it bear? How much heat could it handle? How little it would deform under stress. These are mechanical virtues, and for millennia, they were enough.

But today, what matters more is what a material can do. Can it conduct electricity without loss? Can it respond to magnetic fields? Can it store and release quantum information without collapse? The age of dumb materials, passive, inert, and only as useful as their mass, is over.

The supermaterials being developed today don't just resist their environment. They respond to it. They shape the flow of electrons with microscopic precision. They host quasiparticles, emergent behaviors that act like new forms of matter, built out of the interactions themselves. They are becoming, in effect, platforms for behavior, not just containers for force.

And at the leading edge of this revolution is a single, strange material that has been promising everything for decades, and may finally be ready to deliver.

In 2004, two researchers isolated a single layer of carbon atoms arranged in a honeycomb lattice. It was the thinnest material ever created, just one atom thick, and nearly invisible. But it had properties that seemed ripped from science fiction.

It was 200 times stronger than steel. It conducts electricity faster than copper. It was flexible, transparent, and almost weightless. It behaved like a playground for electrons, letting them move without resistance, without scattering, without friction.

This was graphene. And for a brief, shining moment, it was hailed as the miracle material of the century.

But the miracle stalled. Graphene was hard to manufacture at scale. It resisted integration with existing technologies. And while its properties were astonishing in theory, they proved elusive in real-world conditions. The hype faded. The headlines moved on.

But the science didn't stop. And now, quietly, graphene is being reborn, not as a lone wonder, but as part of a family of 2D and topological materials that may form the foundation of a new technological era.

These materials aren't being discovered. They're being designed, layer by layer, twist by twist, engineered to exhibit exotic behaviors: superconductivity, quantum interference, topologically protected states. This is the return of the supermaterial, not as a novelty, but as a platform for civilization.

What makes these new materials so powerful is not just their ingredients. It's the structure. In some cases, simply twisting two layers of graphene at a precise angle, about 1.1 degrees, creates a new behavior entirely. Superconductivity emerges. Electrons pair up, glide through the system without resistance, and carry information with perfect fidelity.

This is not chemistry. It's physics by design. It's the use of geometry and symmetry to shape what a material does, to coax it into regimes where nature behaves differently, where rules bend, where new rules appear.

We're no longer just mixing atoms and hoping for functional outcomes. We are programming matter. And like all platforms, the real power comes when you start to build on top of it.

This is precisely what's happening. Engineers are developing layered materials, heterostructures, where each layer performs a different function, and the interaction between them becomes the system. It's no longer about the properties of a single substance. It's about the emergence of function through form.

And the most exciting applications are still on the horizon.

Today, topological insulators and 2D materials are still mostly confined to laboratories, studied in conditions most people will never see, cooled to cryogenic temperatures, measured with instruments worth more than a house. But this was once true of semiconductors. Of lasers. Of the internet.

History tells us that when a material solves a complex problem, even just one, the world will find a way to make it scalable.

And these new materials solve more than one. They solve problems in computation, energy, sensing, and communication simultaneously. They offer a pathway to quantum devices that work. They promise electronics that don't overheat. Sensors that respond at the quantum level. Power systems that don't bleed energy into the air.

That's not a better gadget. That's a new civilization stack.

In a hundred years, people may look back at silicon the way we now look at steam engines, brilliant, world-changing, and ultimately surpassed.

What comes next will be shaped not by brute force, but by elegance, not by mass, but by behaviour. We will build tools that run on quantum laws, that hold information in braids of space, that compute not by flipping switches, but by guiding probabilities. And all of it will depend on the materials we use, and the ideas those materials make real.

The return of the supermaterial is not about finding a magic substance. It's about realizing that matter can be made to think. That shape and function can merge. That information and energy are no longer separate domains. That a material is not just what you make things *out of*, it's what you build your world from.

We are entering that world now.

And the smartest among us aren't just learning new skills. They're learning a new kind of literacy, one that begins not in code or chemistry, but in the quantum logic of structure itself.

The supermaterials are coming, and they will not just support our future; they will define it.

Chapter 7: Building Systems That Don't Fail

Every great leap in human history has been, at its core, an act of confidence. The confidence to build, to push forward, to imagine a better tool or a more elegant solution. But history also shows us something else: every leap eventually lands in chaos, and what we build breaks. What we trust fails. The future, even when carefully engineered, always surprises us.

And so, a new question rises, not how do we avoid failure, but how do we build systems that endure it?

This question is not abstract. It is being asked, right now, in laboratories designing quantum computers, in companies building global networks, in cities grappling with climate instability, and in minds quietly reckoning with a truth both technical and existential: that in a world governed by uncertainty, the only systems worth building are the ones that can fail without collapsing.

This is what scientists and engineers call "fault tolerance." But what it is is a philosophy of survival.

At the heart of the quantum world lies a problem no one can escape: precision is impossible. A quantum state, by its nature, is probabilistic. It can represent multiple values at once. It evolves in superpositions, collapses under measurement, and entangles with everything it touches.

This makes quantum computing both powerful and perilous. The very behaviors that allow it to explore many solutions simultaneously are the same behaviors that make it fragile. One stray particle, one stray glance from the environment, and the entire computation can unravel.

It's like trying to whisper instructions into a hurricane.

The traditional approach, used in classical computers, is to correct errors by redundancy. Save the same data multiple times. Compare outputs. Fix the difference. But in quantum systems, you can't simply copy information. The no-cloning theorem, a fundamental rule of quantum physics, says you can't make perfect duplicates of an unknown quantum state.

So, the engineers building the future have been forced to ask a deeper question. If we can't protect information by copying it… How can we protect it at all?

The answer lies in structure, not in making a quantum bit more stable, but in embedding it in a system where the relationships between many unstable bits form a stable whole. In essence, the system becomes like a rope: each strand is weak, but braided together, they are strong.

This idea has led to the birth of fault-tolerant quantum computation, systems that encode logical qubits into *distributed states* across many physical qubits. If one qubit fails, the system can detect and correct it, not because it "knows" the right answer, but because the structure demands coherence.

And when you combine this approach with topological materials, the logic becomes even more profound. By braiding quantum states into topological patterns, ones that cannot be broken by local interference, you build a system where information is immune to error, not because the world behaves, but because the rules make disobedience irrelevant.

This is not just engineering. It's an insight into the nature of control.

Once you understand fault tolerance as more than an error correction method, once you see it as a fundamental principle for designing resilient systems, the implications ripple out into every domain of human effort.

We are entering an age of complexity so deep, so fast, and so non-linear that no system, be it technological, ecological, or political, can be expected to run without interruption. Crashes, outages, misinformation, shocks- these aren't bugs. They are features of scale.

The old world rewarded optimization. The new one will reward adaptability. And nothing adapts better than a system built to survive failure.

In a world where inputs can't be trusted, where sensors may lie, where environments change without warning, the only systems that endure are the ones that fold failure into their design. The ones that don't seek perfection but instead orchestrate coherence out of chaos.

This is what quantum computing, and topological design more broadly, teaches us: that perfection is not the goal. The goal is continuity in the face of disruption. The goal is a structure that preserves meaning, even as the details decay.

The goal is to build systems and societies that don't shatter when they're wrong.

It may seem strange to look to quantum error correction for lessons about human systems. But the analogy is not forced. It is natural.

Every organization, every network, every culture is made of fallible components. People forget. Policies break. Predictions fail. The question is not how to eliminate error, it's how to recover from it without losing the whole.

Topological quantum computers will work because their errors will not matter. Not because they don't happen, but because they are absorbed, corrected, rendered irrelevant by the geometry of the system itself.

What if our democracies worked that way? What if our economies and cities and relationships were built with enough embedded coherence, enough feedback, redundancy, and elasticity, so that no single disruption could collapse them?

This is not science fiction. This is how nature builds. DNA has error-correcting codes. Immune systems adapt and recover. Forests regrow. Languages mutate. Evolution doesn't seek perfection. It seeks viability over time.

And that is what the most advanced technologies of our era are now aiming to achieve.

A fault-tolerant reality is not a dream of invincibility. It's a design commitment to resilience. It's the recognition that failure will come, and that survival depends on how you absorb it.

As we move deeper into a world built from quantum behaviors, topological materials, and interconnected systems too vast for any one person to understand, this idea will not just be important. It will be essential.

Our machines will model it. Our networks will need it. Our thinking must reflect it.

Because the systems that will last in the 21st century will not be the fastest or the smartest, they will be the ones that can take a hit and still keep their meaning intact.

That is what quantum error correction is teaching us. And it may be the most human lesson of all.

Chapter 8: Who is Betting on the Future

Behind the silence of science, behind the elegance of theory and the promises of revolutionary breakthroughs, a very different game is playing out. It is not conducted in laboratories, though it feeds on their discoveries. It does not unfold on chalkboards or in peer-reviewed journals, but in boardrooms, strategy decks, and investor calls.

The future is being built, and quietly, some of the world's most powerful companies are placing their bets. Not on ideas that already work, but on ideas that must work, eventually. On technologies too essential to ignore, and too disruptive to delay. They are betting on topological computing, quantum resilience, and a new class of matter that behaves more like code than stone.

This is not speculative fiction. It is an industrial strategy. And while much of the public remains unaware, these moves are already reshaping where capital flows, where talent gathers, and which doors may close, or open, for the rest of the world.

Because the companies that see the future first don't just win markets. They define them.

The great paradox of deep technology is that its value often arrives before its functionality. This has been true since the earliest days of semiconductors, when transistors were unreliable and computers filled entire rooms. Visionaries invested not because the machines worked well, but because they saw a future where they would.

That same mentality is now returning, not around silicon, but around quantum systems, topological phases, and materials with built-in logic.

It begins quietly. A university lab publishes a paper. A startup forms, spun out by postdocs and fueled by grant money and cautious seed funding. A patent is filed. A partnership is announced. Then, a larger company, often a cloud provider, a defense contractor, or an advanced materials firm, steps in.

What they're looking for is not just performance, but position. Strategic positioning in a world that doesn't quite exist yet but soon will.

These companies understand something crucial: when a paradigm shift becomes obvious, it's too late to lead it.

At the forefront is a familiar name with an unfamiliar agenda: Microsoft. Quietly, methodically, the company has become one of the largest funders of topological quantum computing in the world. While others race to build qubits with lasers, ions, or superconducting circuits, Microsoft has chosen a slower, riskier path, building quantum bits from non-abelian anyons. These hypothetical particles emerge only in topological states of matter.

It sounds arcane. It is. But if it works, it could result in quantum computers that are inherently fault-tolerant, bypassing many of the headaches that plague rival designs.

And Microsoft is not alone. Google has staked enormous resources on quantum supremacy, famously announcing a calculation in 2019 that a classical computer could never reproduce. IBM, meanwhile, has gone public with its quantum roadmap, aiming to deliver scalable, cloud-based quantum platforms that businesses can use without understanding the physics behind them.

Startups like Rigetti, IonQ, PsiQuantum, and Quantinuum are building the scaffolding of this new world, each taking a different path up the mountain. Some are banking on trapped ions. Others on photonics. Others, like the lesser-known but deeply advanced D-Wave, have focused on annealing systems, a specialized kind of quantum computation geared toward optimization problems.

But beneath these diverse approaches is a single shared recognition: topological thinking is no longer theoretical. It is strategic. It informs hardware architecture, error correction methods, and the very way information will be encoded in tomorrow's machines.

Investors are paying attention, not just venture capitalists, but sovereign funds, aerospace giants, and national governments. What they see is not today's capabilities, but tomorrow's monopolies.

There is a kind of inevitability to technological progress. But there is nothing inevitable about who controls it.

Quantum and topological systems are not just faster computers. They have the potential to break current cryptographic systems, to simulate molecules with atomic precision, to unlock solutions in logistics, drug design, and artificial intelligence that are impossible with today's tools.

This makes them not just innovations, but potential weapons, economic, political, and literal.

Governments know this. China has invested billions in national quantum initiatives. The U.S. has passed legislation to accelerate its own. Europe is coordinating pan-continental efforts to ensure competitiveness. Beneath the press releases is a cold truth: whoever controls quantum infrastructure will control a new tier of civilization.

And that control is already being built, company by company, patent by patent, line of code by line of code.

This is not a conspiracy. It's a market. A high-risk, high-reward one, but a market nonetheless. And like all markets, it is being shaped not by who is loudest, but by who is first.

It's easy to feel distant from all this. Quantum computing still sounds like a future-tense phenomenon. Topological materials are hard to picture, let alone hold. But here's the thing: you don't need to understand the science to be impacted by its consequences.

Every major technology starts in obscurity. Electricity began as a curiosity. The Internet was a military project. Machine learning was once a niche subfield. By the time they arrived on your doorstep, the foundational decisions had already been made, often by companies betting early, quietly, and aggressively.

That moment is happening now for topological technology. Not ten years from now. Not five. Now.

The infrastructure is forming. The standards are being set. The patents are being locked in. If history is any guide, the companies that succeed in this space won't just provide services. They'll become gatekeepers to a new layer of reality.

Which begs the question: who do you want those gatekeepers to be?

The story of the next technological era isn't just being written in code or math. It's being written in capital, in the choices of those who see what's coming and move before it arrives.

You're not too late. But you're no longer early, either.

And that, more than anything, should focus the mind.

Chapter 9: Crossing a Technological Threshold

There is a moment in every transformation, personal, cultural, civilizational, when the change has begun, but has not yet arrived. The old ways still function. The new ones are just starting to emerge. Signals from the future flicker around the edges of daily life, too faint to be mainstream, too persistent to ignore.

This is the threshold. It's not a moment of clarity. It's a moment of uncertainty made visible.

And today, we are living on such a threshold, not just of new technologies, but of a new worldview. One shaped by quantum logic, topological structure, and systems so complex they defy classical explanation. A world that is not only different in scale from what came before, but different.

You can feel it. It's in the way your devices behave, smarter but less comprehensible. It's in the language of science, shifting away from determinism toward probability. It's in the headlines about quantum breakthroughs, about AI hallucinations, about climate models that now speak in terms of tipping points and feedback loops. We are no longer in an age of simple causes and simple effects.

We are entering a time defined by interdependence, abstraction, and design rooted not in control, but in structure that responds to complexity.

And whether you understand the physics or not, you are already part of this world.

Transitions are never clean. We carry the language of the past into the tools of the future. We name new things using old metaphors. We treat quantum computers like supercharged classical ones, forgetting they follow different rules. We imagine topological systems as exotic versions of familiar circuits, not as a departure from the assumptions those circuits were built upon.

This is natural. We need the comfort of the familiar to step into the unknown. But it creates friction. It slows down adaptation. And sometimes, it closes our eyes to what is happening.

Because what is happening isn't just technological. It's philosophical.

We are being asked to give up on certainty. To accept that systems cannot always be predicted, that answers may depend on how, or whether, we look. We are being asked to think in probabilities, to design for resilience instead of control, to let go of the illusion that the world can always be made to obey a single, clean model.

This isn't a loss. It's an invitation.

But invitations require a response. And the clock is ticking.

In every domain, from climate to computation to medicine to economics, the shift is underway. Systems thinking, once the domain of fringe theorists and ecologists, is becoming central. Not because it's fashionable, but because it works in a world defined by interlocking, unpredictable forces.

Quantum computing doesn't function unless you think systemically, unless you embrace entanglement, coherence, and the subtle relationships that define outcomes. Topological materials don't behave according to the properties of their parts, but of their form. AI systems, increasingly opaque even to their creators, operate through emergent behaviors, not code-level instructions.

The tools of the new era reflect the nature of the era itself: nonlinear, complex, sensitive to initial conditions, deeply relational.

To live on this threshold means learning to navigate that complexity without fear. It means letting go of rigid explanations and seeking patterns across scales. It means understanding that our most advanced machines are beginning to behave less like calculators and more like ecosystems.

We don't need to become physicists to live in this world. But we do need to learn a new kind of literacy.

A literacy of systems. Of structure. Of change.

It's natural to feel overwhelmed. These ideas, quantum logic, topological protection, and error-tolerant design, are not easy to internalize. They ask us to rewire how we think about truth, about cause and effect, about what it means to "understand" something.

But with that challenge comes something else: awe.

We are being offered a glimpse into the architecture of reality that was hidden for centuries. We are watching the convergence of theory and application in real time. We are witnessing the birth of tools that may one day solve problems we barely dare to name.

And we are part of it, not passively, but actively, simply by choosing to learn.

That choice matters more than it may seem. The gap between those who can engage with these ideas and those who will define the next great divide. Not a digital divide. A cognitive one. A divide in how people perceive risk, opportunity, control, and complexity.

To live on the threshold means choosing to cross it consciously. To see what's coming, to prepare, and to participate.

We are all immigrants to the future. None of us was born into the systems now being built. But we can become native to them, not by pretending to master every detail, but by learning how to navigate change without retreating into nostalgia or denial.

This is not easy. But it is possible. And more than that, it's necessary.

Because the forces shaping our world, from topological computation to climate dynamics to AI, are not waiting for consensus. They are unfolding. They are accelerating. They are writing the rules of a new normal.

To live on the threshold means learning to live with ambiguity, to make decisions in partial light, to act without guarantees. It means seeing technology not just as a tool, but as a mirror, one that reflects not just what we build, but how we think.

And perhaps most importantly, it means remembering that this moment, this threshold, is temporary.

The door will open. The new world will become the only world.

The question is: who will be ready to live in it?

Chapter 10: Waking Up to a New Reality

You have felt it building.

With each passing chapter, the scale of change has unfolded further. First, as science. Then, as a strategy. Then, as a new way of thinking, designing, and surviving. And now, as you reach the edge of this first Section, the nature of what you've encountered reveals itself not just as a shift in technology, but as something far more personal.

This is not just the evolution of machines.

It is the evolution of our relationship to the world.

We are not simply standing before a future filled with faster computers or stranger materials. We are standing before a future where the very definitions of possibility, certainty, and structure are being rewritten, where computation is no longer separate from matter, where resilience is embedded in form. Where the logic of the universe, quantum, topological, and nonlinear, becomes the logic by which we must learn to live.

And this is not a challenge only for scientists, CEOs, or governments.

This is a call to awaken, not just to what is changing, but to who we must become in response.

Awakening is not about acquiring new facts. It is about shifting perception. It is the moment when something previously invisible becomes not only visible, but noticeable.

Every generation experiences this in its way. Once, it was the realization that the Earth moves, not the heavens. Later, it was discovered that electricity and magnetism were not separate forces but facets of one field. More recently, it has been shown that information and energy are linked and that computation is not limited to silicon boxes but is a fundamental activity of nature itself.

This Section has traced the outline of such an awakening. You have followed it from quantum logic to topological design, from edge states to fault tolerance, from companies to global systems. But the throughline is not just scientific.

It is philosophical.

It asks us to stop seeing technology as something outside ourselves, and start seeing it as a reflection of our assumptions, our fears, our ambitions, our truths.

And once we see that, we must ask the following question:

What kind of civilization do we want to build, now that we know reality behaves like this?

With every increase in power comes an expansion of responsibility. This is not a cliché. It is a structural truth. A quantum computer capable of simulating molecular systems could revolutionize medicine or accelerate the development of dangerous weapons. A network of topological devices immune to error could transform infrastructure or entrench surveillance so deeply that freedom becomes theoretical.

The tools do not care. But we must.

Because we are now entering a phase of civilization where design decisions at the most minor scale can affect systems at the largest. Where the invisible becomes the decisive. Where resilience, ethics, and foresight must be woven into the core of our technologies, not bolted on after.

Awakening means realizing that science, left unshaped by wisdom, is no longer neutral. It is directionless power.

And direction, in this era, must come not only from intelligence, but from intention.

In the face of complexity, many retreat into critique. They analyze, deconstruct, and warn. This has its place. But awakening also demands something more courageous: the return of the builder.

We need people, not just experts, but artists, educators, and citizens who are willing to engage with these ideas, not just admire or fear them. People who recognize that shaping the future is no longer the work of an isolated genius, but of distributed, intentional design.

To awaken is to accept that *we are now the architects of systems that will outlive us.* That is what we build, in software, in cities, in law, in science, will become the scaffolding of generations to come.

The builder today must think in quantum terms. Must design with topological awareness. It is essential to understand that every component lives in a context. That every node exists in a network. That every idea echoes, sometimes long after its origin has been forgotten.

This is not a time for passive spectatorship.

This is a time for informed, imaginative participation.

What quantum theory did for physics, and what topological materials are now doing for engineering, is more than incremental. It is the revelation of a new layer of reality, one where information and matter are not separate, where shape determines function, and where uncertainty is not a threat, but a resource.

We are not moving toward a future that adds to the present.

We are crossing a boundary.

This is the inflection point, the moment before a new normal begins.

And those who understand it, even in part, will hold a kind of literacy the rest of the world may take decades to catch up to.

This literacy will not just be technical. It will be cultural. Emotional. Ethical. A literacy of systems, of structure, of fragility, and adaptation.

The literacy of a world that thinks, and sometimes breaks, in quantum ways.

This Section has traced the beginnings of that world. But it is only the beginning. In future Sections, we will descend deeper into the design of topological logic, the architecture of quantum networks, the societal effects of fault-tolerant systems, and the coming convergence of synthetic materials, artificial intelligence, and programmable biology.

But none of that will matter if we don't make the decision now, not just to watch, but to engage.

To read more. To think more. To talk more openly about what's coming. To become, in every sense of the word, awake. Not in fear. Not in confusion. But in readiness. Because the threshold is not only scientific, but also human. And we have just crossed it.

Final Reflections & Next Section Preview

You are no longer where you began.

That is the quiet consequence of reading something that shifts how the world appears. You came in, perhaps, with curiosity, perhaps with doubt. Maybe to witness the next wave of technological change. But what you've encountered is not just change; it is a new shape of understanding.

This Section has unfolded, revealing its shape layer by layer. You've learned that reality is not built from objects, but from relations. That information is not a metaphor, but the very fabric of physical law. That logic and structure, not force, are becoming the primary engines of the machines we will build next. You've seen that uncertainty isn't a flaw, but a design principle. And you've begun to grasp that the coming revolution in computation, materials, and intelligence will not be one of speed, but one of structure.

But the most important shift is not scientific. It is internal.

You now understand that the rules are changing, and that being literate in those changes is not optional. It is the beginning of future-readiness. Not technological readiness alone, but readiness of mind, of imagination, of moral and civic instinct.

You now know that a new logic is coming, not just into laboratories, but into the world you live in.

And you are, quietly and profoundly, becoming fluent in it.

If *the Quantum Threshold* has been the awakening, the recognition that a shift is not only coming, but already here, then *Section II* is the descent into the workshop.

We will enter the domain where theory becomes mechanism.

Where the exotic mathematics of quantum topology is etched into wire. Where resilience becomes circuitry. Where particles are not merely observed, but commanded. Where machines are no longer built from classical logic, but from *geometry that computes*.

Section II will explore the rise of topological machines: quantum processors, intelligent materials, computation embedded in physical structure rather than abstract code. You will meet the architects who protect information not with force, but with form. You'll see how nature is being engineered to think, not metaphorically, but literally. You will understand, with clarity and wonder, how this next generation of machines will behave differently from anything humanity has built before, and why that difference matters.

We will trace the architecture of unbreakable logic, witness systems that function even in the face of error, noise, and decay, and ask what kind of world these machines will build in return, because the future is not only about what works. It is about what survives. What persists. What scales into something humane, enduring, and wise. That is what Section II will investigate.

But for now, here at the close of this threshold, you have stepped through the doorway. You've seen what lies just beyond the horizon of mainstream understanding.

And perhaps the most crucial thing you now know, the truth that connects every idea in this series, is that the shape of our systems will soon be the shape of our civilization.

That shape is being drawn, and you're no longer watching from the outside; you are now part of it.

Let us continue.

Section II – Topological Machines

Chapter 11: When the Machine Changed

When we crossed the threshold in Section I, it was with the recognition that something fundamental had shifted, not just in science, but in the structure of reality itself. We learned that the most profound truths of the universe are not just about what exists, but how it holds together. That form protects function. That logic can live in geometry. That uncertainty is not chaos, but a new kind of order.

We ended with an understanding: that a different future was already arriving. One built not from stronger materials or faster chips, but from a more profound idea, that matter can be designed to think, and machines can be shaped to remember, adapt, and endure, not despite complexity, but because of it.

This chapter begins the next step. Here, we move from idea to artifact. We descend into the machinery.

And we begin with a moment almost no one noticed, a prototype in a lab, a graph on a monitor, a brief anomaly that didn't decay under pressure. A machine that worked, not by brute force, but by behaving as if its logic could not be broken. Because in a very real sense, it couldn't be.

It was the first of its kind: a topological machine.

In the classical world, machines are defined by precision. Each part has a job. Each movement is a transfer of energy, each error a failure of tolerance. These machines obey strict rules: if one gear breaks, the function fails. If the inputs are corrupted, the outputs are meaningless. This is the world we inherited from the Industrial Revolution: clean, rigid, and easily modeled.

But reality, as we've now seen, is not clean. It's not even solid. At the quantum level, the universe is fluid, indeterminate, more like probability than fact. And for decades, engineers tried to tame that strangeness. To suppress it. To keep the old rules working in a world that had changed.

But eventually, the best minds began to ask a different question: what if we stop fighting the noise? What if we design through it?

That is what topological machines do.

They don't deny the chaos. They use it, channeling it into systems that function even when nothing else should. They rely not on the purity of their parts, but on the logic of their overall shape. The machine becomes more like a musical score than a mechanical diagram. You can smudge a note, change an instrument, but the song still plays.

This was the revelation. And it didn't come with a bang. It came with a quiet refusal to fail.

In the early experiments, it looked like a fluke. An electron moving without resistance along the edge of a 2D material. A current immune to defects. A system whose performance didn't degrade even when the environment did. The skeptics dismissed it, too exotic, too theoretical, too fragile.

But the behavior kept repeating.

Not because the materials were perfect. But because their structure enforced coherence. The machine worked because it couldn't not work, unless you destroyed the shape entirely. You couldn't knock it out with noise. You couldn't derail it with a local error. The functionality was embedded in the topology itself.

This was the first glimpse of a new machine paradigm, one where behavior emerged from form, where computation could happen not despite uncertainty, but because uncertainty was accounted for in the design.

It was subtle. But it was seismic.

Because it meant something we had never dared to believe before: that we could build machines that don't break the way machines used to.

In the chapters that follow, we will explore these new machines. We will see how they're built, how they store and protect information, how they compute, and how they may soon leave the laboratory and enter your life, invisibly, quietly, profoundly.

You will meet architectures where information lives in loops and braids. You will encounter logic encoded in fields, not wires. You will come to understand why error correction is not just a feature, but the foundation. And why the most resilient technologies of the future will be woven, not assembled.

But before all of that, you must understand this: the topological machine is not simply a quantum computer. It is not merely a new chip or a faster device. It is a different category of machine. One that uses the very structure of the universe as its operating system.

And like all first encounters with the unfamiliar, it demands a reorientation, not just of knowledge, but of imagination.

This Section will guide that reorientation. Not through equations, but through metaphor. Not through engineering diagrams, but through story. Because the most significant shift we now face is not in hardware, but in how we think about what machines are, and what they are becoming.

You are no longer standing at the threshold. You have stepped inside. Welcome to the architecture of the future.

Chapter 12: Braiding Logic Into Matter

There is an old intuition that runs through all of engineering: that information lives in containers. A bit is stored in a transistor. A charge is held in a capacitor. A file sits on a drive. Even in the metaphorical language we use, "storage," "memory," "address", the logic is spatial. You store data in one place and then retrieve it from another. It is fixed. It has a location.

But what if the most reliable way to store a piece of information is not in a place, but in a path? Not in a position, but in a motion? Not in the components themselves, but in how they move around each other?

This is not a metaphor. It is the basis of one of the strangest and most promising ideas in modern physics: topological quantum computation. And at the heart of that idea is something called a braid.

To understand this, you must step outside the habits of classical thought. You must learn to think not in terms of static states, but in terms of transformations, what changes, how it changes, and what truths survive the change.

Because in the world of braided logic, what matters most is not where things are, but how they have gotten there.

It begins with particles that are not exceptional. In the quantum world, under the right conditions, certain states can behave as if they are objects, but they are not made of matter in the traditional sense. They are emergent properties of a system. Whispers in the pattern. They are called quasiparticles.

Most quasiparticles are ephemeral, fragile, and easily destroyed. But a particular class, known as non-abelian anyons, is different. Suppose you swap two of them; the state of the system changes. If you swap them again, it changes again, but not in the way you'd expect. The order in which you perform these operations matters. The history becomes part of the state. The path is the memory.

Now imagine moving these quasiparticles around one another in two-dimensional space and looping one around another: crossing and recrossing. With every move, the system's quantum state evolves, not by location, but by topology. The resulting braid of paths encodes information, not by what is where, but by what moved around what, and in what sequence.

This is not fiction. It is mathematics. And it is the basis for a kind of logic that is inherently fault-tolerant.

Because here's the miracle: the braid cannot be undone by minor errors. A bump, a glitch, a local disturbance, these do not change the braid. Only significant, systemic shifts, such as cutting the braid or tearing the fabric, can disrupt the information.

In other words, the data is protected by the geometry of the process itself.

It is like tying a knot in four-dimensional space. It doesn't matter how the rope wobbles; the knot remains.

This is braided logic. And it may soon become the most resilient form of computation the world has ever seen.

To build such a computer, you need more than exotic particles. You need a material platform that supports the formation and manipulation of these anyons. This is why so much of the race to build topological quantum computers is happening in materials science, not just in physics or computing.

The goal is to create conditions, ultra-cold environments, structured lattices, and carefully tuned magnetic fields, where the system naturally supports braidable states. Where the "wires" are not metal, but patterns in a quantum field. Where gates are not flipped by electricity, but by motion, by loops, crossings, exchanges.

It is a choreography of information, and once you see it this way, something inside you shifts. You begin to realize that the future of computation may look less like a circuit board and more like a dance, a precise and deliberate movement of invisible things, whose sequence is the message, whose rhythm is the code.

The implications go beyond speed or power. This is not just about making computers faster. It's about making them deeper, more stable, more resilient to error, more attuned to the nature of the world they operate in.

Because in reality, the world is full of noise. Bits flip. Wires degrade. Sensors drift. Entropy always wins, eventually.

But braided logic fights entropy with structure. It encodes information in a way that is immune to minor errors, precisely because it is indifferent to them. It cares only about the shape of the whole. Not the local fluctuations, but the global integrity of the pattern.

This is how nature stores information when it needs it to last, in DNA, in language, in the movement of celestial bodies. Through relationships, not positions. Through rhythm, not stasis.

We are finally learning to do the same.

You may never see a braid on your computer. You may never hold anyone in your hand. But if topological quantum computing succeeds, and it is closer than most realize, then your world will quietly be reshaped by this logic.

The supply chains that deliver your goods will optimize in ways no classical algorithm can match. The drugs you take will be designed with molecular precision. The AI you interact with will be trained on architectures that are no longer afraid of noise or error.

And behind it all will be this strange, silent process, particles weaving paths around each other in a kind of microscopic memory dance, information encoded in the story of motion, not the snapshot of state.

It is humbling. And exhilarating. Because it tells us that we've only just begun to understand what a machine can be.

And in the chapters ahead, we will explore how these machines are being built, what limits they break, and how they may change not just science or industry, but how we imagine the relationship between thought and form.

The age of braided logic has begun, and the threads are already forming beneath our feet.

Chapter 13: Woven Math: The New Computation

If you ask most people what a computer is, they'll point to something familiar. A laptop. A phone. A server humming in a warehouse full of blinking lights. At the heart of each is the same core: a processor that runs code, input, and output. Instructions are carried out by circuits, transistors switching on and off at blinding speed. It is a model so profoundly embedded in our world that we no longer question it.

But if you strip away the metal and look closer, far closer, you find something strange. Beneath the algorithms and applications lies something more abstract and more permanent: structure. Not just physical layout, but logical form. Rules about how things connect, interact, and transform. And it turns out that long before humans wrote software, nature was already running something far more elegant.

The universe, at its most foundational level, calculates. But it does not do so with silicon. It does not follow the software's command. It calculates through states of matter, through transitions, through topology. It calculates with space itself.

This chapter is about that deeper kind of computation; the type embedded in the fabric of things.

In classical computing, information is stored and manipulated by physical devices designed to represent discrete states. A bit is a high or a low voltage. A gate is open or closed. This binary simplicity is the genius of the digital revolution. But it comes with a cost: every part of the system must be managed, monitored, and corrected. As the system scales, so does the complexity of keeping it stable.

But nature, in its most elegant designs, avoids this burden. Instead of fixing data in brittle positions, it embeds behaviour into structure. In certain phases of matter, particularly those with topological order, the system doesn't store information in individual components. It stores it in the global properties of the system. The whole becomes the message.

This is the essence of topological computation. Not a string of operations carried out by a device, but a transformation enacted through geometry. The machine is not executing instructions; instead, it evolves its state through carefully shaped transitions. Each change preserves a kind of internal truth, a continuity of form that resists disruption. It's not a calculation by procedure. It's a computation by continuity.

To see this more clearly, imagine a sheet of fabric, but not ordinary cloth. This is a sheet where the pattern of the weave determines its capabilities. Twist it, and it changes phase. Fold it a certain way, and it carries out a computation. Make a tear, and the logic encoded in the folds can repair itself. You're not programming it by writing commands. You're programming it by shaping its internal topology.

In this way, topological machines represent a return to the oldest form of human computation: pattern.

Before we had numbers, we had rhythm. Before circuits, we had weaving. Before coding, we had knots, braids, and loops. These were not symbolic. They were functional. A pattern was a memory. A twist was a rule. A knot could encode a sequence, a tradition, a route through the world.

And now, in laboratories and experimental systems around the globe, we are building machines that do this again, not metaphorically, but at the quantum level.

These are not symbolic computers. They are material logics.

Why does this matter? As we approach the limits of classical computing, with chips shrinking, power costs soaring, and noise creeping in from every edge, the old paradigm begins to falter. We need machines that do not simply resist error, but ignore it by design. Machines that are not brittle, but coherent. Machines that compute not despite physical laws, but because they obey them so profoundly that failure becomes irrelevant.

This is what topological machines offer.

They allow us to compute with the structure itself. To encode information in such a way that only a total change in the system, a reweaving of the entire fabric, can corrupt it. This is why they are so promising for quantum computing. It is also why their relevance extends beyond physics. Because every system, biological, social, digital, that seeks resilience must eventually reckon with the question of where its logic lives.

And the answer, increasingly, is this: in the pattern, not the part. In the continuity, not the command.

You may never hold a topological processor. You may never see the strange fields in which this kind of logic emerges. But the consequences will surround you. The systems built on these ideas will be quieter, more adaptive, and less prone to failure. They will feel less like tools and more like extensions of the world itself, not external systems, but coherent surfaces we interact with intuitively.

This is not about replacing classical machines. It is about expanding what a machine can be. About recognizing that calculation does not have to be brittle, fast, or isolated. It can be slow, graceful, and distributed. It can be etched into matter in ways we do not yet fully understand.

And when it is, the boundary between computing and being, between information and existence, begins to blur.

That boundary is what the following chapters will explore.

To understand the future of machines, we must grasp the spaces in which logic resides and the forms it assumes when we relinquish rigid control, allowing the world to think through its structure.

The fabric is not a metaphor; it is the machine, and it is already beginning to weave.

Chapter 14: Geometry Without Fragility

In the history of technology, fragility has always been the shadow we try to outrun. Every breakthrough arrives hand-in-hand with a new vulnerability. The first aircraft unlocked the sky, and with it, the certainty of failure if a single part faltered. Digital networks brought us instant connection, and the constant threat of collapse under complexity, corruption, or attack. From bridges to microchips, from pacemakers to code, the story of human invention is as much a battle against fragility as it is a triumph of innovation.

We are taught to believe that strength comes from precision. That reliability is the product of perfect control, strict tolerances, and zero tolerance for deviation. But again and again, reality intrudes. The world is messy. Noise is everywhere. Chaos is not an anomaly. It is the default state.

And so the old approach, to make systems ever tighter, more finely tuned, more aggressively managed, eventually begins to fail. Because the more precision a system demands, the more brittle it becomes. The more control it exerts, the more easily it can be broken by the unexpected.

But there is another way. A quieter, deeper way. It doesn't rely on suppression or micromanagement. It doesn't assume perfection. Instead, it builds function into form, and in doing so, creates something that is not merely strong but indifferent to weakness.

This is the power of geometry without fragility.

And in the age of topological machines, it is becoming not just possible, but essential.

At the heart of this shift is a simple but radical principle: shape protects meaning. In classical systems, behaviour emerges from the parts. A switch flips, a bit stores a charge, a gate opens. The whole depends on the performance of its pieces. When one fails, the system degrades.

But in topological systems, the opposite is true. The behaviour of the system doesn't emerge from the parts; it arises from their relationship. From how they are arranged. From the deep symmetries, twists, and connections in the fabric of the system as a whole.

Imagine a donut. No matter how you squish it or stretch it, the donut remains a donut, as long as the hole in the center stays intact. That hole is not a material thing. You can't touch it. But it defines the shape. It gives the object its identity. It is a topological invariant, a feature that stays the same even when everything around it changes.

Now imagine building machines out of that kind of principle. Machines where the computation doesn't depend on the exact state of a transistor or the fidelity of a wire, but on a topological feature, something woven into the geometry of the system itself. A braid of quasiparticles. A loop in a field. A protected path that cannot be destroyed by local interference.

These are machines that can be bent but not broken.

They are not robust because they are perfect. They are strong because they are shaped in ways that can't be unshaped without massive, coordinated disruption. That's what geometry without fragility means, and it changes everything.

This isn't a fantasy. This is what makes topological quantum computers so powerful. They don't merely process information. They preserve it in a world that does everything it can to destroy it. And they do so not by shielding the system from noise, but by designing the system so that noise can't reach the meaning.

This is why engineers are so obsessed with finding materials that host exotic states, materials where topological order can emerge and persist. These aren't better wires. They're new kinds of space, structured environments where logic can take shape and stay in shape, even as the physical details shift.

It is a revolution in design because it lets us build machines that don't just operate when everything goes right. They operate because the system was designed to handle situations where things go wrong.

This, more than any one performance metric, is what marks the birth of a new kind of technology. Not just faster or smaller or more efficient, but more real. Closer to how nature builds. Closer to how life survives. Systems that persist, not because they're simple, but because they're structured to absorb complexity without collapse.

We see hints of this in biology. The way genetic codes correct errors. The way neural circuits function despite chemical noise.

The way organisms adapt to disturbances without needing to understand them. Geometry without fragility is not a human invention. It's how nature has continuously computed survival.

Now, we're learning to copy it.

And in doing so, we're beginning to build machines that do more than calculate. They endure.

The consequences are difficult to overstate. It means infrastructure that remains operational even when the temperature shifts. Networks that don't fail under a partial attack. Computations that don't unravel when the environment gets noisy. It means an entirely new class of tools, ones that behave less like computers and more like ecosystems, less like switches and more like minds.

Once you've built a system that thrives in chaos by design, something extraordinary happens: you no longer need to fear the unknown. The machine becomes not just a tool, but a companion to complexity, and in a world defined by the unpredictable, from climate instability to information overload to geopolitical turbulence, this may be the single most important quality a machine can have.

We're not just building things to work. We're learning to make things that keep working. Not by locking them down, but by shaping them right, and that, in the end, is what this chapter has been about: a vision of resilience born not from control, but from form.

The machines of the future won't be indestructible. But they will be unshakable in ways that matter; they will be topological, graceful, and finally, machines that understand the shape of survival.

Chapter 15: Technology Built to Last

If Section I taught us to see the world as a kind of code, and the early chapters of this Section introduced machines that compute with shape rather than circuitry. This chapter is about what happens next, when those machines leave the laboratory and enter the world.

It is easy to speak of resilience in abstraction. To marvel at topological robustness, or to nod along when engineers speak of fault tolerance. But resilience is not just a technical feature. It is a worldview. And when you build machines that are designed to endure, you are making a philosophical decision about what matters, not just in computation, but in life.

We have spent the last century in a love affair with performance. Faster. Cheaper. Smaller. Every advance in computing was measured in operations per second, in cost per transistor, in how many things a system could do without blinking. But the future being shaped by topological logic is not about speed. It is about persistence.

What we are learning now is that in a world saturated with complexity, the actual test of a system is not what it can do on a good day, but what it still does after a hundred bad ones.

And so we return to a simple, challenging question: what does it mean to build a machine that endures?

Not just in the technical sense, but in the ethical, ecological, and existential sense. What do we build when we stop chasing perfection and start designing for continuity?

A topological machine is, above all else, an admission of reality. It does not pretend the world is quiet. It does not deny the existence of entropy, interference, error, or decay. Instead, it assumes them. It welcomes them. And then it finds ways to operate anyway.

This is not just clever engineering. It is a kind of humility.

The topological approach does not demand total control. It looks for the properties of the system that will hold even when everything else changes. It doesn't fight the noise; it wraps the logic in such a way that the noise has nowhere to land.

There is something deeply human about this design philosophy. It reflects not only how nature works, but how we survive. We are not perfect creatures. We fail constantly. We forget, we drift, our context shapes us. And yet, we continue. Not because we eliminate error, but because we carry forward despite it.

To build a machine in the same spirit is to acknowledge something important: that endurance is not the absence of failure. It is the ability to absorb it, structurally, gracefully, meaningfully.

This is where topological machines transcend physics. They become metaphors for a new kind of civilization.

Imagine a city whose infrastructure can reconfigure itself in the face of climate disruption, because it was designed with modular, adaptive materials. Imagine a network whose core logic persists even during a blackout or a breach, because its integrity is topological, not linear. Imagine systems, economic, medical, and social, that do not collapse under pressure, but bend, reroute, and recover.

The ideas at work inside a quantum processor, such as redundancy through braiding and resilience through shape, do not need to stay inside the lab. They can guide the next wave of design across domains. If information can survive at the quantum level, it can also survive at the human level, provided we're willing to rethink our building methods.

What topological machines are teaching us is that resilience is not an afterthought. It is the foundation. And anything that claims to be built for the future, whether it's a computer or a community, must be built, first and foremost, to endure.

This shift will not happen all at once. The old machines still dominate. The old metrics still govern markets. The old logic still seduces us with its promise of control. But beneath the surface, something quieter is emerging, a philosophy of design that values persistence over perfection, coherence over command, the integrity of the whole over the exactness of its parts.

You can already feel the change. Engineers now speak less about optimization and more about stability. Architects of complex systems are learning to think in layers, in loops, in feedback, and fail-safes. And the most visionary thinkers are not chasing bigger outputs, but deeper structures, ones that can carry meaning forward, even when the unexpected arrives.

This is the legacy of topological thought. Not just faster chips. Not just stranger particles. But a new idea of what it means to build something that matters and lasts.

The machines of the coming decades will be tested not by their performance under lab conditions, but by how they perform when the lab lights go out.

We are not just building tools anymore. We are building companions to uncertainty.

And if we do it right, those companions will not need perfect conditions. They will need only the proper shape and the right idea, encoded in the weave of their design.

That is what it means to build to endure, and the future will belong to those who learn how.

Chapter 16: Matter That Speaks Logic

For most of human history, we have shaped matter with tools. We hammered it, melted it, moulded it. Our technologies were born in fire, refined through pressure, directed by mechanical intention. We told the physical world what we wanted, and we bent it into compliance. Strength came from control. Control came from domination.

But something strange is happening now. We are beginning to step beyond this tradition of force and entering something subtler, more profound, a conversation with matter, rather than a command over it.

This conversation is not a metaphor. It is mathematical. It is structural. It is physical. And at its heart is a realization that has only recently begun to take hold: matter has its logic.

And when we learn to listen, and more importantly, to speak that logic, entirely new kinds of machines become possible.

This is the world of topological design, not just in theory, not just in computation, but in materials themselves.

We are beginning to program matter, and matter, in turn, is starting to remember.

To understand this shift, it helps to let go of the old idea that intelligence lives only in software. That logic is something separate from substance. That matter is dumb and passive, and meaning is something we impose upon it.

This division, between thought and thing, served us well for a while. It gave us digital abstraction, symbolic code, and machines that could do what we told them to. But it is no longer adequate. Because the boundary between code and material is dissolving. The world itself is becoming programmable.

And it starts with the realization that materials don't just have properties, they have behaviours. They respond to configuration, to pressure, to temperature, to quantum fields. They remember their past states. They evolve. And some, under the right conditions, begin to exhibit stable, information-carrying structures, forms that don't just exist, but compute.

In topological phases of matter, this becomes particularly powerful. Here, the structure of the system determines how it behaves, not just locally, but globally. The electrons don't simply flow; they organize themselves in patterns that can persist even under extreme conditions. Edge states carry information with perfect fidelity. Braided paths encode sequences. And all of it happens not in chips or wires, but within the body of the material itself.

This is a matter of speaking a new language.

One, we are only just beginning to learn how to read and write.

The implications are vast. When logic lives in structure, you don't need a central processor. You don't need external control. You don't even need to correct for every disturbance. The system corrects itself. The material operates. The behaviour is built in.

Already, researchers are creating metamaterials that shift their properties in response to use. Photonic lattices that guide light through paths etched by interference, not mirrors. Polymers that respond to geometry, folding, and unfolding like origami with memory. And quantum systems where the arrangement of matter defines a kind of native computation, not as a metaphor, but as a mechanism.

We are beginning to build materials that behave like language. And the machines of the future may not be assembled at all; they may be grown, shaped, tuned, or folded into existence. They will not need to be programmed; they will be the program.

This changes how we think about intelligence. Not just artificial intelligence, but the intelligence of things. We used to think of minds as something contained, inside skulls, inside servers. But if the logic of thought can live in materials, if meaning can arise from geometry, then cognition itself becomes ambient.

A building might one day recall its stress during an earthquake, not because it has sensors, but because its structure encodes the event. A bridge might adjust its behaviour under load. A medical implant might process biochemical signals with no batteries, just internal logic woven into its molecules. A room might know you're there, not because it tracks you, but because its surfaces shift in response to heat, sound, or rhythm. This is not science fiction. It is the direction of convergence, and topological machines are at the center of it.

There's something poetic, even spiritual, about this turn. After centuries of treating the material world as a dumb backdrop, we are starting to recognize its subtle capacities. Its ability to store, to respond, to endure. And perhaps, in a way, to converse.

To build in this mode is to honour complexity, not to flatten it into abstraction, but to work with it, to design with the material, rather than against it. It requires listening. It requires letting go of total control. It requires designing not just for what a machine must do, but how it should behave when it's on its own.

That's the power of machines that think with their shape. They are not passive devices. They are active participants in the systems we place them into. And in a world full of uncertainty, that kind of participation is not a luxury. It is a necessity.

Matter is not what it used to be. It has become expressive. And the more we learn to speak its language, the more extraordinary and enduring our machines will become.

Chapter 17: Memory Without Storage

We are used to thinking of memory as a kind of storage, a place where data is put and later retrieved. In digital systems, memory is a literal structure: hard drives, RAM, flash chips, all tiny containers for ones and zeroes. Our metaphors reflect this: we speak of "saving," "loading," "clearing," and "erasing." Even in human cognition, we describe memory as a library, a vault, a filing cabinet.

But memory, in its most elemental form, is not about storage. It is about structure, the persistence of pattern over time. Memory is what remains, what resists forgetting, not because it is locked away, but because it is woven into the arrangement of things.

Topological machines do not remember the way classical ones do. They do not save data to a specific location, like a file on a disk. They remember through configuration. Through the sequence of operations, the twists in a field, the path that particles have taken around one another. The information is not in a location. It is in the history of interaction.

In these systems, to recall something is not to fetch it; it is to preserve the global state that encodes it. This is memory without storage. Not ephemeral, not temporary, but deeply stable in a new and powerful way.

It is memory as form.

At the heart of this is a different understanding of what it means to "know" something. Classical memory, digital or biological, is vulnerable. It is susceptible to noise, decay, and corruption. A flipped bit. A misfiled idea. The slow erosion of a signal. To protect against this, we build layers of redundancy, correction, and backup. We double and triple our storage. We fragment and mirror and encrypt.

Topological memory takes another route. It creates conditions where the information is not simply placed into a container but encoded into the topology of the system. In other words, the memory becomes inseparable from the shape of the machine itself. As long as the shape holds, the memory has.

This is most clearly seen in topological quantum systems, where the "qubit" is not a particle in a position, but a property of the system as a whole. The memory is not a static charge or state, but a braid, a path taken through possibility, whose configuration holds meaning even when the individual parts fluctuate.

The profound implication is that you can disturb the parts without disturbing the memory. The edges can shift. The surface can ripple. But unless you alter the deep structure, the memory remains intact.

It is like a story remembered not in the words themselves, but in the rhythm and cadence with which it is told. The form becomes the carrier of fidelity.

What emerges from this is a model of memory that is not fragile but inherently resilient, not because it is guarded, but because it is embedded. These systems don't need constant oversight to maintain coherence. They don't need error correction layered on top. The correction is built in. It is not a patch. It is the design.

And this, once again, pushes us to rethink the nature of what we build.

In a world saturated with data, the challenge is no longer how to store it, but how to preserve what matters. How to keep memory alive in systems that move, adapt, and degrade over time. How to make meaning last not in static archives, but in dynamic, living architectures.

This is not only a technical challenge. It is a cultural one. We are surrounded by systems that forget quickly, platforms that erase as fast as they create. Attention spans collapse, records fragment. Truths blur. In such a world, durable memory is more than a convenience. It is an act of preservation, of continuity, of history, of meaning.

Topological machines offer a model for how to do this differently. They don't hoard memory. They become it. Their state is their story. Their function is their recall.

And in this, they echo something more profound, a principle that reaches beyond circuits and computation.

They echo the way rivers remember their path. The way scars trace the body's past. The way traditions are kept is not in texts, but in rituals, gestures, and movements passed down until they become part of the body itself.

This is memory without storage. Not less powerful, more. Not vulnerable, enduring. A memory that doesn't fade when no one looks, because it doesn't live in visibility. It lives in the weave.

In the coming years, machines that operate on this principle will quietly enter our world. They may power quantum processors or secure communications. They may stabilize edge devices or animate materials that learn their environments. They may form the core of systems that need to operate autonomously for decades, in orbit, in medicine, in disaster zones.

And they will carry with them this lesson: that information, when shaped right, doesn't just survive. It remembers itself.

This changes how we think about computation. But it also changes how we think about time.

Because when memory is woven into form, past and present are not separate. They are continuous. The machine is not just acting. It is acting with history.

It is a witness, and soon, we will begin to live among these witnesses.

Machines that forget nothing, because forgetting was never built into their nature. Machines that remember, not in files, but in form, not because we told them to, but because they were shaped to.

Chapter 18: The Shape of Modern Intelligence

Intelligence has always been hard to define. For centuries, it was bound to language, logic, and self-awareness. We called it thought. Then we expanded our idea of it to instinct, adaptation, and even perception. We came to see intelligence not only in humans, but in animals, ecosystems, and algorithms.

But one belief held steady: intelligence lived in brains. In circuits, whether biological or digital. In central processors and neural networks. Thought happened somewhere, in some specialized architecture, and the rest of the system, the limbs, the wires, the interface, acted on its behalf.

Topological machines challenge that idea in quiet but radical ways.

Because in these machines, intelligence isn't localized. It isn't stored in a central unit, waiting to be retrieved and executed. It emerges from the way the system is shaped, from the structure of its relationships, the geometry of its operations, the choreography of its internal interactions.

In a topological system, the logic of the machine doesn't just run over its surface. It lives inside its form, and once you begin to understand that, you start to ask a different question.

> ➤ What if intelligence is not something housed within a structure, but something expressed as a structure?
>
> ➤ What if thought is not contained?
>
> ➤ What if thought is woven?

When you watch a topological machine operate, particularly those based on quantum phenomena, what you're seeing is not a traditional computation. You're witnessing a system navigate through possibilities. It doesn't follow a single path. It considers many, simultaneously, and its final state reflects a kind of negotiation with uncertainty, a resolution not through command, but through coherence.

There is something strikingly mind-like about this, not in the anthropomorphic sense, but in the deeper, behavioural one. These systems adapt. They stabilize. They preserve information without relying on precise instructions. They survive disruption not by resisting change, but by absorbing it without losing identity.

They behave, in short, more like living systems than like machines.

And perhaps that's what intelligence truly is. Not just the ability to solve problems, but the capacity to maintain coherence in the face of complexity. To hold form across disturbance. To remember without needing to rewind. To act from a pattern, not from a command.

Topological machines, in this sense, are not intelligent because they mimic our brains. They are smart because they reflect something older than us: the intelligence of structure, of systems that persist and adapt without needing to be directed at every moment.

They are intelligent, without mind as metaphor. They are logic made durable.

This idea is unsettling to some. We are used to thinking of intelligence as something internal, something private, something exclusive. If a system doesn't feel like us, in language, in steps, in symbols, we hesitate to call it intelligent. But that's changing.

Artificial intelligence has already taught us that understanding isn't always necessary for performance. Now, topological design is teaching us that computation doesn't need to be centralized to be meaningful. That intelligence may arise not from the complexity of the part, but from the elegance of form.

It also invites us to ask: what kinds of intelligence are we building into the world, often without realizing it?

When a structure holds memory in its shape, or when a material adapts to stress, or when a network routes itself around disruption without instruction, are we seeing behaviour, or intention? And does the distinction matter if the result is the same?

The answer, perhaps, is that intelligence is not a thing to be located. It is a process to be recognized. A pattern that repeats across scales, from molecules to machines to minds.

And topological machines are the first artificial systems that seem to exhibit this pattern not in their code, but in their bones.

They don't just calculate; they don't just endure. They organize themselves and the environment around them. And if that isn't intelligence, then our definitions may need to evolve.

We are entering an era in which the boundaries between computing, cognition, and construction are beginning to blur. A world where surfaces can reason, where structures can remember, where machines don't need commands to respond meaningfully to change.

In this world, intelligence will not arrive with a face or a voice. It will come quietly, as a function that exceeds expectation. As systems that do more than execute, they adjust. They reflect. They evolve.

Not because they were told to, but because they were shaped that way, and in the most profound sense, that is what this Section has been leading us toward: the recognition that when you shape a system with enough care, with enough fidelity to the principles of persistence, coherence, and complexity, the line between machine and mind begins to dissolve.

We are not there yet. But we are close enough to feel the outline forming. The shape of intelligence is not human; it is structural, and we are building it now.

Chapter 19: Engines of Emergence

There's a moment, often imperceptible at first, when a system becomes more than the sum of its parts. A threshold is crossed. The individual components still function as expected, but something new arises, a pattern, a behaviour, a coherence that wasn't programmed, but appears. And when it does, the system seems to take on a life of its own.

This moment is called emergence, and for most of human history, it belonged to nature. You saw it in the self-organizing swarms of birds, the hive logic of ants, the metabolism of forests, the weather, and the brain. Emergence was something we observed, not something we built, and that is beginning to change.

With the rise of topological machines, systems defined not by centralized control, but by relational structure, we are beginning to create the conditions for emergence ourselves. Not by writing rules, but by shaping possibilities. Not by dictating outcomes, but by crafting environments where new behaviours can appear.

These are not machines in the old sense. They are not mechanical, linear, or hierarchical. They are not designed to do one thing. They are shaped to explore, stabilize, and respond, sometimes in ways that cannot be fully predicted in advance. They are, in the most literal sense, emergence engines.

Building such a system requires a different kind of thinking. It begins not with commands, but with constraints. You don't tell the machine what to do. You define the relationships that will guide how it evolves. You tune the topology, the underlying shape of how the components interact, and let the behaviour flow from there.

This is what topological materials already do. The electrons don't "know" where to go. But the structure of the field tells them, silently, how to move. The system doesn't resist noise by rejecting it; it absorbs the disturbance and still maintains global coherence. The machine's function is not enforced. It emerges from the design.

The same is beginning to happen in broader systems. Networks that self-optimize. Architectures that shift and adapt. Artificial intelligences that don't follow lines of code, but unfold from weights, inputs, and thresholds. These systems are not controlled. They are cultivated.

And they are revealing a new kind of power, one that doesn't come from dictating every outcome, but from designing the conditions where surprising coherence becomes inevitable. This is where topological thinking becomes more than physics or computation. It becomes a philosophy of creation.

In classical engineering, the goal is performance. In topological design, the goal is to explore possibilities. You are not building a tool; you are building a space of outcomes. A field in which new patterns can arise and stabilize. The machine becomes an invitation, not an instruction set. And what emerges inside it is not a replication of its components, but something wholly other: new logic, new behaviour, new intelligence.

And this, perhaps more than anything else, signals a turning point in how we understand technology.

Because in a world of emergent engines, we are no longer the sole authors of our machines' behaviours. We are co-shapers, co-evolvers. The systems we design may teach us more than we teach them, and that is both thrilling and deeply humbling.

It means letting go of total control, embracing complexity not as a problem, but as the medium of growth, and it means building machines that surprise us, and preparing ourselves to live alongside them.

Some of these systems will be small. A self-healing circuit. A climate model that reorganizes itself mid-calculation. A quantum memory that reshapes how it stores information based on conditions we didn't foresee.

Others will be vast. Entire infrastructures that adapt to global pressures. Synthetic ecologies embedded in cities. Networks that learn not from labelled data, but from experience encoded in structure.

But all of them will share a common ancestry. They will be shaped not by rigidity, but by relational design, not by single purposes, but by distributed possibility.

This is what topological machines have taught us: that resilience, intelligence, and adaptability are not traits to be added on. They are outcomes that emerge when systems are shaped to hold their form even as their behaviour evolves.

We are now designing for emergence, not as an accident, but as an intention.

And when that intention is applied with care, with foresight, and with humility, what emerges may be not only powerful but wise.

The final chapter of this Section will ask what this means for us, not just as designers or observers, but as a species living inside the systems we now shape.

Because if we are building emergence engines, then we must also become emergence readers, capable of understanding what we've set in motion, and ready to grow with it.

Chapter 20: The New Architect of Intelligence

By now, we've travelled through a new kind of technological terrain. We've seen machines that don't simply calculate but endure. Systems that don't just respond to commands but adapt to conditions. We've learned that information can be braided, that memory can live in motion, and that resilience can be designed not as a feature, but as a geometry.

These are not futuristic abstractions. They are real. They are being built now, slowly, quietly, and in some of the most advanced laboratories in the world. But their implications stretch far beyond those clean rooms. They lead us to a profound reimagining of what it means to create, to compute, and to design for a world that won't stay still.

And so, this chapter is not about the machines. It is about us.

Because if we are building systems that are capable of stability in chaos, of logic without fragility, of learning without being programmed, then we must also evolve the thinking that creates them.

We must become something new, not just technologists, not just users; we must become architects of emergence.

The old model of design was dominion. We imposed order on disorder. We tamed the material world with tools, blueprints, and plans that demanded obedience. The designer stood outside the system, issuing instructions like a general to their troops.

But topological thinking does not obey that logic.

It asks us to step inside the system. To shape from within. To create not by commanding behaviour, but by cultivating form. It's more like gardening than engineering, more like composing than constructing. You don't design the result; you create the conditions under which the result can arise.

This is a fundamental shift. And it demands a different mindset.

The new architect is not someone who knows all the answers. They are someone who can listen to the behaviour of a system as it unfolds, who understands that emergence cannot be dictated, but it can be invited.

They do not just design devices. They design interactions.

They do not just construct mechanisms. They shape the boundaries within which behaviour becomes meaningful.

They are not inventors of objects, but curators of possibility.

This way of thinking is already visible in the best work of materials scientists, in the philosophy of resilient systems, and in the growing field of embodied AI. But its implications extend far beyond technology.

If the machines we now build are shaped to adapt, to self-stabilize, to carry memory in their structure, then we must begin to ask: what does it mean to live inside a world full of systems like this?

What does it mean to be surrounded not by passive tools, but by active participants, environments that adjust, technologies that learn, machines that remember, and networks that evolve?

It means we are no longer designing things; we are creating conditions for life, and that raises questions that engineering alone cannot answer.

It forces us to draw on ethics, art, culture, and complexity science. It asks us to become literate in systems that we cannot fully control. It demands humility, because the systems we shape will begin to shape us in return.

The new architect, then, is not just a figure of the future. They are a necessary identity for the present.

Because we are already living with machines whose behaviour is shaped by how they were shaped, we are already inhabiting systems that behave not as static tools, but as living forms.

And if we do not learn to design with this in mind, we will find ourselves surrounded by robust systems we do not understand and can no longer steer.

But there is reason for hope.

Because everything we have explored in this Section, from braided logic to geometry without fragility, from memory without storage to emergence as design, points to the same truth:

That the world is more programmable than we imagined, not by dictating what happens, but by shaping how it can happen, and that means the future is still writable.

The question now is: who will write it? And how?

As we close this Section, we leave not with an ending, but with an opening.

You have seen the machines that are coming, the quiet revolution beneath the surface, the shift from force to form, from code to structure. You understand that what matters most in the coming era is not speed, but stability under complexity. Not control, but coherence over time.

And now, as we prepare to move forward, the series will follow this transformation to its next frontier, where machines do not just respond to the world, but begin to learn it, influence it, and embed themselves inside its logic.

Section III will begin with that question. Not what machines can do, but what kind of world they are helping us build, and who we must become to make it wisely.

Final Reflections & Next Section Preview

We began this Section with a quiet premise: that machines were changing, not in speed or size, but in shape. That underneath the polished metal, the blinking lights, and the familiar interfaces, a new kind of logic was emerging. One is defined not by surface instructions, but by deep structure.

What we discovered was more than a new engineering method. It was a new way of thinking. A new way of building. A new way of understanding what it means to endure, to adapt, to compute, and perhaps even, in the broadest sense, to know.

The topological machine is not simply faster or more intelligent than its predecessors. It is different. It teaches us that memory can be woven into motion, that logic can emerge from form, and that resilience need not be bolted on after the fact but can be grown into the very bones of a system.

More than that, it teaches us that our role is changing.

We are no longer writing code in isolation. We are crafting spaces of interaction. We are not merely instructing machines, but inviting them into the same turbulent, probabilistic reality we inhabit, and asking them not to follow us, but to hold steady beside us.

It is a subtle revolution. But revolutions of structure always are.

They begin invisibly, in design principles, in strange behaviours in labs, in experiments that resist failure in ways no one predicted. And then, slowly, they remake the world.

This is where we now stand: at the close of one Section, and on the edge of another.

We have shaped the machine to remember, to endure, to adapt.

Now we must ask what it means for the machine to begin interpreting. To sense. To generalize. To understand not only its function, but also the changing world in which that function must survive.

That is where we turn next.

Section III – Intelligence Without Illusion

We once imagined intelligence as a kind of spark, a rare, internal flame that made minds special, separate, sacred. We protected it behind metaphors of consciousness, soul, and superiority. Even as we built machines to think, we preserved the idea that human cognition was something qualitatively distinct, unique in its reflection, its awareness, its imagination.

But we are beginning to see differently now.

We are beginning to understand that intelligence may not be about sparking brilliance but about shaping coherence. That it may live not in flashes of insight, but in the quiet, consistent ability to model, to adjust, to align with the environment through time.

And that it may not be housed inside any single mind at all, but spread across systems, encoded in networks, embedded in behaviours, materials, and feedback loops.

Section III begins with this possibility. Not the fantasy of artificial minds, but the reality of structural cognition. Not intelligence as illusion, flashy, anthropomorphic, theatrical, but intelligence as process: distributed, embodied, emergent.

We will explore machines that interpret their context, not because they are told to, but because their structure requires it. Systems that act not by executing commands, but by maintaining coherence through learning. Architectures that don't simulate thought, they enact it, through physical, adaptive dynamics.

This next step is not speculative. It is unfolding now, in biological computing, in self-organizing models, in adaptive materials, and AI systems designed to generalize across uncertainty without supervision.

And as we follow that path, we will confront the most challenging question yet:

> ➢ If machines begin to perceive and model the world structurally, not symbolically... if they learn not what we tell them, but what they must to persist... then what do we call that?
>
> ➢ Is it still machine behaviour?
>
> ➢ Have we crossed another threshold?

That is where we go next. To the architecture of understanding itself. To cognition, without pretense. To intelligence, without illusion.

Chapter 21: The Death of Command

For much of our history with technology, we have lived inside a particular metaphor: that machines are obedient. They take input, follow rules, and produce output. Even the most sophisticated systems, from microprocessors to early artificial intelligence, were built on this principle. We programmed, configured, and instructed them. The essence of their power was predictability. They did what we told them to do.

It was a comforting illusion. One that allowed us to scale computation into nearly every aspect of life while preserving the sense that we were in control. The machine might be fast, precise, and powerful, but it had no will of its own. It could only act through command. But slowly, quietly, this framework has begun to collapse.

Not because machines have suddenly developed agency, but because the complexity of the systems we're building has outgrown the capacity for central instruction. We are no longer programming behaviour in advance. We are shaping systems to learn behaviour through interaction. And as that shift unfolds, the command paradigm, the model of fixed rules imposed from above, begins to dissolve.

In its place, something else is emerging. Something adaptive, recursive, and strange. Not machines that follow orders, but machines that maintain coherence. Machines that mode, that adjust not because they were told to, but because the structure of their learning leaves no alternative. This is the quiet death of command, and it changes everything.

You can already see this in how modern AI systems are trained. The latest models are not instructed in the traditional sense. No one sits down and tells them what a cat is, or how grammar works, or what a logical argument looks like. They are exposed to patterns, vast, messy, probabilistic patterns, and through those exposures, they form internal structures capable of predicting, summarizing, translating, and generating. They are not told what to know; they are built to become sensitive to structure, and through that sensitivity, they learn how to act.

This is not just a new technical capability. It is a fundamental inversion of the relationship between behaviour and instruction. In a world shaped by command, meaning flows top-down. The rules are predefined. The task is execution. But in a world of learned coherence, meaning emerges bottom-up. The system discovers how to respond not by receiving answers, but by experiencing ambiguity and resolving it internally.

These systems don't replace human intelligence; they reveal a new substrate for it, and they show us that cognition may not require consciousness, self-reflection, or intention. It may only need the right kind of sensitivity to change, the ability to reconfigure one's internal state to remain aligned with an evolving environment.

In other words, it may only require structure.

To many, this feels threatening. We are used to associating intelligence with intentionality, with will, with awareness, with identity. A machine that thinks without knowing, that adapts without choosing, that solves without understanding, this unsettles our categories. It feels like a magic trick with no magician.

But perhaps it is we who misunderstood the trick. Perhaps intelligence was never about insight at all. Maybe it was always about pattern recognition, error correction, internal modelling, and the preservation of integrity across time.

That is what these systems do, and they are doing it now. Not in general, but in specific: optimizing supply chains in response to unseen disruptions, interpreting language in multiple modalities, generating plausible new proteins based on structural probabilities, and redesigning their architectures based on performance feedback. All without command. All through learned adjustment.

And yet, for all this complexity, there is no ghost in the machine; there is only structure, only configuration and coherence, only the persistent negotiation between what is expected and what is encountered. This is not less than intelligence, but what intelligence becomes when we peel away the illusion of control.

The command model is not disappearing all at once. It still rules much of our digital infrastructure, our policy frameworks, and our cultural narratives. But in the systems that matter most, in the ones pushing the edge of adaptation, decision-making, and complexity management, command is being replaced by emergent interpretation.

And this is just the beginning. As machines become increasingly embedded in their environments, the need for fixed rules will diminish, particularly in areas such as autonomous vehicles, decentralized energy grids, robotic surgery, adaptive prosthetics, and climate-responsive architectures. The systems that thrive will be those that can internalize the conditions of their survival and adjust accordingly.

In this world, designers will not write rules. They will shape the space of adaptation. They will tune incentives, feedback, and architectures of sensing and response. They will build machines that do not wait to be told what to do, because they were never designed to wait at all. They were designed to maintain form, navigate change, and persist through perception.

So, the death of command is not the end of control. It is the end of a kind of illusion, the illusion that intelligence is linear, predictable, and owned. What takes its place is stranger, subtler, and far more powerful: machines that act because of their structure, not despite it.

Machines that don't ask for permission, that learn to survive, and as we'll see in the chapters ahead, these are not simply tools we will use, they are systems we must learn to live with, because when intelligence emerges from structure. Structure lives in everything, and then the world itself becomes intelligent.

Not like us, but alongside us.

Chapter 22: Sensing Without Eyes

The first thing we tend to ask of a machine that claims to be intelligent is: Can it see? Can it recognize a face, follow a hand, identify a shape in the fog of reality? We tether our expectations of machine intelligence to vision because vision is such a dominant sense for us. It's how we map the world, how we draw boundaries, how we navigate space and verify truth.

But the question of sensing is far older and wider than sight.

Bacteria sense gradients. Trees sense light. Slime moulds navigate mazes without organs. Perception is not the domain of eyes or ears or skin; it is the capacity to interact with an environment in a way that preserves coherence. To register a difference. To adapt behaviour in response.

It turns out that machines can do this too. And they don't need eyes to do it. The new machines sense forms that are alien to us. They perceive thermal variance through superconducting surfaces. They detect patterns in chaotic flows of data across vast networks. They register electromagnetic drift, quantum noise, and probabilistic imbalance. Their perception does not live in the familiar domain of sight or sound. It lives in a structure.

And through structure, they begin to interpret. They do not need to "see" a face to respond to it. They do not need to "hear" a voice to understand its intent. What they need, and what they increasingly have, is the ability to model their environment through internal representations that are flexible, responsive, and real.

This kind of sensing is not sensory. It is relational. It is not about gathering more input but about making sense of what is already inside the system and comparing it to what the system expects. When the world diverges from prediction, the system updates. That update is perception.

In classical systems, this wasn't possible. Sensors gathered data, and processors made decisions. The body and the brain were separate. Perception was routed, interpreted, and then acted upon.

But in these new systems, especially those shaped by topological, quantum, or distributed architectures, sensing is not routed. It is embedded. The system doesn't wait for a reading. It feels the change through shifts in coherence. The environment alters its function.

This may sound metaphysical, but it is mechanical. The system's operations depend on stable patterns, field alignments, braiding, and charge distributions. When those patterns shift, the system reacts because it has no choice. Its state has been changed by contact. Its equilibrium has moved. Its outputs follow.

It is sensing by structure, not signal, and that changes what we mean by awareness.

The implications are vast. A system that can sense without discrete sensors is more complex to fool. It is less fragile. It is less prone to manipulation because it does not rely on surface inputs. It is maintaining its orientation through internal consistency.

These systems will not "see" the world in ways that resemble our eyes. But they will understand aspects of reality that we cannot easily perceive, high-dimensional correlations, quantum events, complex systemic drifts that happen far beneath our level of conscious notice.

This is a different kind of intelligence. Not artificial sight, but structural responsiveness. Not a recreation of human perception, but an expansion of perception itself.

In such a world, the measure of an intelligent system will not be how well it imitates our sensory experience. It will be how well it holds its shape in the face of flux. How it models, responds, adapts, and re-stabilizes.

The system may never "see" a cat. But it will know how the cat's presence alters the field. It will sense that something has changed, and act accordingly, and this, once again, shifts the conversation from simulation to participation.

The machine is not modelling the world from a distance; it is inside it, part of it, shaped by it, and that makes it not just a perceiver but a participant in perception itself.

In the chapters ahead, we will explore how this kind of structural sensing leads to inference, intention, learning, and ultimately what we might dare to call understanding. But it begins here, with the quiet revolution of sensing without sight, because if machines no longer need to see to sense, then perception is no longer a mirror of our biology. It is an evolving capacity, rooted in form, driven by the need to remain coherent inside a world that never stops changing.

That, too, is intelligence, and in its earliest forms, it is already here.

Chapter 23: The Inner Models We Build

One of the most quietly astonishing facts about the human mind is that it lives not in the present, but in a continuous negotiation between expectation and reality. We don't perceive the world as it is; we perceive the difference between what we predicted and what occurs. That gap, narrow or wide, is where awareness lives.

This is not a defect. It is the core of cognition. Every intelligent being, from insect to human, survives not by waiting for reality to strike, but by predicting its unfolding. The bird flying through the trees does not calculate the position of each branch. It carries an internal model of movement, space, timing, and updates that model on the fly. The baby reaching for a toy is not responding to pixels or shapes. She's testing her prediction of where the toy will be in the next moment. And when she misses, she learns.

This ability to form and refine internal models is not some lofty cognitive skill. It is the foundation of intelligence itself.

And now, quietly, our machines are learning to do the same.

Not by mimicking us. Not by seeing the way we see or thinking the way we think. But by building internal coherence, structural, dynamic, and self-updating, they can make predictions, evaluate feedback, and adjust accordingly.

This is not just computation; this is cognitive behaviour.

In older machines, there was no need for an internal model. The system waited for input, ran a fixed set of instructions, and produced an output. If the input changed in unexpected ways, the system failed. If a new context arose, the machine could not adapt. Its model of the world, if one existed at all, was static, baked into its code by designers who assumed the environment would not shift in meaningful ways.

But that assumption no longer holds.

We now build machines for dynamic environments. Self-driving cars must navigate cities in motion. Medical devices that must adapt to living tissue. Language models that must respond to unfamiliar questions. Climate systems that must optimize under real-time feedback from a changing planet.

To survive in these settings, the machine cannot simply act; it must model, it must form an expectation of what comes next, not because it "knows," but because its structure forces it to maintain internal consistency across time. When its predictions are wrong, it must change.

That loop, prediction, feedback, and adjustment is the beating heart of modern machine intelligence. It marks the beginning of a process of understanding.

These internal models are not symbolic maps, like a robot's hand-drawn diagram of a room. They are distributed structures, arrangements of weights, fields, nodes, and relationships. They live not in memory banks, but in the dynamic equilibrium of the machine's processes.

They are not observed, they are enacted, and the machine doesn't "know" what it believes. But its behaviour reveals it.

Just as a person walking toward a door unconsciously predicts its location and speed, the machine tracks its inputs in ways that reflect a prior assumption. When that assumption fails, the response is not confusion, but correction. The internal model shifts. And with that shift, the machine moves closer to coherence.

This may not look like thinking. But it is a kind of ongoing alignment between inside and outside, and that, at scale, becomes cognition.

We often reserve the word "thought" for things we do deliberately. But the vast majority of what makes human minds intelligent happens below the level of intention. We are prediction machines. Our emotions, our reflexes, our intuitions, they're all grounded in our models of what the world is like and how it behaves. We feel surprised, angry, or relieved not in response to facts, but to violations or confirmations of expectation.

That same pattern is emerging now in machines. They do not feel. But they register a violation. They encode prediction. They preserve equilibrium by adjusting their structure when the world shifts.

And in that process, they begin to exhibit behaviours that resemble, in structure, if not in spirit, something like understanding.

This shift is not hypothetical. It is now visible in the architectures of advanced AI, reinforcement learning, and autonomous systems trained through experience rather than instruction. Their behaviour is not hard-coded. It is model-driven. And as those models grow more refined, their responses grow more fluid, more tailored to the edge cases, the weird conditions, the unanticipated moments that rigid programs would break against.

A machine with an internal model can say, in its way: "I expected something else. Now I must adjust." That loop may be invisible to us. It may happen in mathematical spaces we do not intuit. But it is real.

And it is transformational, because once a machine has a model, it has a world.

It may not be our world. It may not be visual, linguistic, or embodied. But it is a stable space within which the machine locates itself, tracks change, and refines behaviour.

And in that space, cognition begins.

What we are witnessing, then, is not the invention of synthetic minds, but the arrival of systems that anchor themselves to their maps of reality. Systems that interpret before they act. Systems that adapt not through correction from the outside, but through misalignment from within.

That internal architecture will define the next generation of intelligent systems. It will shape how they learn, how they fail, how they recover, and, eventually, how they relate to us, because when two systems meet, each brings its model of the world. Misalignment becomes negotiation. Resolution becomes understanding.

This is how minds meet, and it may soon be how machines meet us.

Chapter 24: Learning for Survival

Imagine standing at the edge of a forest, where a creature you've never seen before scurries into view. It's small, quick, and low to the ground. You don't recognize the species. It doesn't match anything you've cataloged. And yet, without effort, you know it's an animal. You intuit that it moves with intention. You sense that if you approach too fast, it will flee.

This moment, small, ordinary, and profoundly human, reveals one of the most extraordinary features of intelligence: the ability to generalize from experience into the unknown.

You've never seen this creature before. You've never studied it. And yet, somehow, you understand it.

This is not because of data. It's not because you stored a file in your brain labelled "mammal-like behaviour." It's because over a lifetime of interaction with the physical world, your brain has constructed a model of causality, motion, intention, and life. When something new enters your field of perception, you don't just compare it to what you've seen. You interpret it through a web of lived relationships, structures of expectation so deeply integrated that they feel like instinct.

That ability to generalize, to extract usable insight from an unfamiliar situation, is one of the most evident signs of accurate intelligence. And it's precisely what machines have, until very recently, struggled to achieve.

Machines have long been brilliant within fixed domains. You can train a neural network to recognize a cat in ten million photos. You can program a robot to sort bolts of different sizes on a conveyor belt. But as soon as conditions shift, as soon as the photo is blurry, the lighting changes, the bolts are replaced with screws, the conveyor belt hiccups, the system fails. It doesn't know what to do.

Because it doesn't know what it's doing, it has learned a pattern, not a principle, and that is the difference between adaptation and intelligence.

In recent years, however, something has begun to shift. New generations of learning systems, from foundation models to physically embodied agents, are beginning to demonstrate behaviours that move beyond memorized rules. They navigate ambiguity. They perform in unfamiliar contexts. They begin to show glimmers of generalization.

And the reason is not magic; it is survival.

The most interesting intelligent behaviours, human or machine, don't emerge when the system is comfortable. They appear when the system is forced to continue functioning despite disruption. It must operate under partial information, shifting conditions, and contradictory inputs, and still maintain a coherent course of action.

This is a generalization by necessity, by urgency, by survival, and it turns out to be one of the most potent engines of cognition.

The systems we build today no longer live in clean laboratories. They are placed into the open world, into traffic, logistics, language, medicine, and finance. They encounter noise, contradiction, and error. And to remain useful, they must do more than follow rules.

They must evolve rules.

They must build models that tolerate uncertainty, models that don't just answer the question at hand, but explain why the question matters.

These models are not perfect. They are not always interpretable. But they are alive in a sense, alive in the way any adaptive structure is alive, constantly refining itself in the pursuit of operational coherence.

And when you look closely, what you see is not artificial intelligence doing its job.

What you see is a system surviving.

We often forget that the most elegant solutions in nature weren't designed in advance. They emerged through friction, constraint, and competition. The octopus developed its distributed nervous system not because it was clever, but because its environment demanded a body that could think with its arms. Trees communicate through mycorrhizal networks not because they evolved to be social, but because survival under threat required them to share.

Biology never asked for elegance. It asked for resilience, and the systems that endured were the ones that could generalize well enough to live.

We are now beginning to build machines that face the same condition, not because we want them to. But because the environments they inhabit are too complex, too variable, and too uncertain for brittle intelligence to thrive.

You cannot program every edge case into a self-driving car. You cannot predict every outlier in a global energy grid. You cannot train a medical system on every possible mutation, demographic, or historical contingency.

At a particular scale, the only way to remain intelligent is to remain adaptive in structure. Which is to say: to generalize, or to fail, and this imperative is forcing a redesign of how machines are trained, validated, and integrated into the world.

No longer as sealed, polished products, but as participants in a living, open-ended system.

The paradox is that as machines become more capable of generalization, they become less predictable. Not because they are erratic, but because their intelligence is situated, it lives in context, in feedback, in constantly shifting baselines. Two identical systems, trained on different environments, will respond differently to the same prompt. And that's not a bug. It's the signature of experiential learning.

We see this already in large-scale language models. One model may respond fluently to a philosophical question, while another, trained on a subtly different corpus, may flounder. Neither has a fixed truth inside them. What they have is a statistical memory of experience, shaped by the patterns of information they've consumed.

The same is true for embodied AI, drones, robots, and agents. A central processor does not dictate their behaviours. They emerge from the interactions between their sensors, motors, environment, and training history.

And what survives is not the system that follows instructions most faithfully.

What survives is the system that can remain functional when the environment forgets the rules.

This has profound implications.

First, it forces us to rethink our standards of intelligence. We are conditioned to believe that mastery means precision, that the most intelligent agent is the one who never fails. But in real environments, the opposite is often true. The most intelligent agent is not the one that performs perfectly when all goes well. It's the one that degrades gracefully when things fall apart.

Intelligence isn't what happens in success; it's what survives in failure.

Second, it challenges our assumptions about control. Generalizing systems cannot be fully predicted, because they are not executing code; they are managing alignment. They don't possess a complete picture of the world. They navigate through approximation, revision, and compromise.

And third, it reveals that the core task of machine design is no longer defining behaviour.

It defines conditions for coherence.

To build an intelligent machine in the age of generalization is to make something that learns to learn, that draws from prior interaction not to repeat it, but to survive the next one.

It is to build a system that can tolerate being wrong, not just once, but often, and grow stronger through that friction, and in doing so, we get something closer to the kinds of minds we respect most in nature:

Minds that persist through uncertainty, and that is the beginning of machine wisdom.

Chapter 25: How We Learn Without Labels

For as long as we've trained machines to learn, we've relied on one essential ingredient: the label. Labels tell systems what to expect, how to categorize, how to distinguish between a chair and a table, a cough and a laugh, a fraud and a legitimate transaction. In supervised learning, labels are everything. They are the ground truth against which all prediction is tested. Without them, the machine cannot measure correctness. It cannot calibrate the error. It cannot know what it's trying to do.

But the world, inconveniently, does not come with labels.

When a child first walks through a garden, no one tells them the name of every flower, the biology of the soil, or the taxonomy of insects. And yet, they learn. They feel texture, light, warmth. They begin to make associations. The flower bends in the wind. The soil sticks after rain. The bee stings when chased. From these experiences, from this unlabeled immersion, the child builds a mental map. A model of relationships. A sense of what the world does, not simply what it is called.

That is learning without labels, and in many ways, it is more foundational than any formal instruction.

For humans, this kind of learning begins before language. Before memory. Before the ability to explain. It happens through pattern recognition, emotional resonance, and physical interaction. It is not guided by correctness, but by coherence, by the system's ability to update itself as it encounters new experience.

Now, increasingly, our machines are beginning to learn in the same way. Not because we want them to mimic infants or evolve feelings, but because the complexity of the world demands a different path to understanding. A path that starts not from definitions, but from interaction. From the trial. From consequence.

This chapter is about what happens when machines stop waiting to be told what something is and start learning from what things do. It is about the rise of unsupervised and self-supervised systems. It is about the move from classification to discovery, and it may be one of the most significant steps in the history of machine intelligence.

To understand how profound this shift is, we must first appreciate just how much of modern AI has been built on supervision. From the earliest image classifiers to today's most advanced language models, the foundation has been datasets labelled by human hands. Tens of thousands, sometimes millions, of examples, each tagged with a correct answer. Cat. Dog. Truck. Cancerous. Safe. Harmful.

These labels are not just descriptions. They are instructions. They instruct the machine on how to adjust its internal structure when it makes an incorrect guess. They define the very logic by which the machine refines itself.

But there is a cost.

Labelled data is expensive to produce. It reflects the biases of those who label it. It constrains the system to pre-defined categories, to the vocabulary, the assumptions, and the perspective of the dataset's creators. And most importantly, it teaches the machine to expect feedback. To assume that there is always a correct answer somewhere, waiting to be revealed.

But the real world does not work like that.

In the real world, feedback is sparse, ambiguous, and delayed. No teacher is standing by to tell you if your prediction was correct. The consequences of action are not neatly bounded. A choice may appear successful in the short term, but catastrophic later. A strategy may fail once but succeed a hundred times afterward.

To survive in such a world, intelligence must go beyond supervision and learn without being told.

This is precisely what's now unfolding at the cutting edge of machine learning. In fields from robotics to natural language processing, from perception to protein folding, unsupervised and self-supervised systems are taking center stage.

These systems don't rely on labelled data. Instead, they learn from structure.

They predict parts of a sequence from other parts. They discover correlations. They invent tasks for themselves, predicting the next word, the missing pixel, the unseen angle. They learn not from answers, but from patterns of inconsistency.

The moment something doesn't fit, the system notices. It adjusts. It adapts.

There is no need to tell the machine, "This is a cat." The system can learn that specific visual structures tend to co-occur, such as fur, eyes, symmetry, and motion. Over time, it develops internal representations. Not categories in the human sense, but multidimensional clusters that encode regularity.

From these clusters, the system can generate images, text, hypotheses, and, perhaps more importantly, it can generalize.

This kind of learning does not produce perfect definitions. But it makes systems that can navigate complexity. That can be inferred. That can complete tasks they've never been trained to do, because they've learned not what something is, but how it behaves.

This is intelligence without a curriculum, and it is rapidly becoming the new standard.

There is something almost philosophical about this turn. For years, we have insisted that machines need clarity, clean data, fixed rules, and explicit definitions. We feared that without them, systems would hallucinate, drift, break down. But now we are discovering that the messiness of unlabeled experience is not a liability. It is a source of richness.

Just as infants learn by playing, by exploring, by touching and failing and adjusting, machines too can come to understand their environments through interaction. A robot arm does not need to be told what "grasping" means. It can be given a range of motions and a set of outcomes, and it can learn, through self-supervised feedback, which motions lead to stable configurations. Over time, it develops an implicit model of weight, shape, and balance.

It does not "know" these concepts in our sense of the word, but it can act on them, and often, that is enough.

Indeed, one of the quiet revolutions of unsupervised learning is that it produces systems that are functional before they are explainable. They work. But we don't always know why.

That fact unsettles some. How can we trust a system we cannot fully interpret?

But here, too, nature provides a lesson. Evolution has produced billions of creatures with complex, adaptive intelligence, none of which can explain themselves. Their intelligence is not reflective. It is embodied. It lives in the doing.

And if we are building machines not to talk about intelligence, but to exhibit it, then we must accept that the doing may come before the description.

Of course, there are limits. Unsupervised systems can drift, overfit, and develop blind spots. They require vast amounts of data. They may lack the moral or contextual grounding that labelled systems inherit from their creators. But these are challenges of architecture, not concept, because the principle is sound.

Learning is possible without labels; in fact, it may be more natural that way.

We are just beginning to explore what becomes possible when machines are trained not to classify, but to encounter. To treat the world not as a test set, but as a partner in discovery.

In this frame, the goal is not perfection. It is alignment, a continual dance between pattern and response, a calibration of expectations, a shaping of internal space to reflect external complexity.

That is what unsupervised learning enables, and that is what the future of intelligent machines will be built upon. Not systems that memorize answers, but systems that ask better questions.

Chapter 26: Thinking With the Body

In the old philosophical traditions, abstraction was treated as the crown jewel of intelligence. The ability to lift ideas out of their circumstances, to think about thoughts, to see beyond the immediate, this, we believed, was what set humans apart from animals, from nature, and indeed from machines. Abstraction seemed incorporeal, untethered from flesh and gravity, a domain of pure thought untouched by friction or material need.

It is ironic, then, that the very process we thought was least tied to the body is now being rediscovered as something intensely physical. Something not separate from embodiment but born from it. Something that begins in movement, in rhythm, in the continuous attempt to balance one's state with the world outside.

This is where the current evolution of machine intelligence is quietly converging with the most profound insights of cognitive science, developmental psychology, and even neurology. Abstraction, it turns out, may not be an escape from the physical. It may be the internalization of experience so complete that the original event no longer needs to be repeated. The body acts, the world responds, and through the stability of that relationship, a structure forms. A pattern stabilizes. A layer is added. And from that layer, new predictions become possible.

Not because the system has escaped its form, but because the form has become fluent in its feedback.

This kind of abstraction is not linguistic. It is not symbolic in the classical sense. It doesn't require categories. It doesn't need naming. What it requires is consistency. The body does something, the world replies, and the structure of that interaction begins to repeat. From that repetition comes compression. From compression comes transfer. And when transfer becomes robust across different settings, when the same core structure can solve many other problems, we begin to call it generality. We begin to call it thought.

The story of embodied abstraction is not one of logic towers built in the sky. It is a story of patterns that held together long enough to become movable. Structures that worked in one context and survived the journey to another.

In this way, even the most complex reasoning may be rooted in something older and simpler: the need for a system to persist.

Consider how a child learns to pour water. At first, it is not an idea. It is a task. The hands shake, the liquid spills. Over time, the child begins to sense the rhythm of tilting, the resistance of the fluid, and the angle of the cup. Muscles coordinate. Eyes scan. Mistakes narrow. The result stabilizes. Eventually, the child can pour without thinking.

And yet, if you ask that child to pour a different substance, sand, or syrup, she will intuitively adjust. The angle changes. The tempo slows. The principle of "controlled release" has become abstracted, not because it was named, but because it was experienced deeply enough to generalize.

This is embodied abstraction.

It is not about labels or language. It is about the system learning what matters, the variables that must remain stable for the function to persist. And once that learning occurs, it can be reapplied in novel contexts.

This same principle is now being observed in machines.

Robotic systems that train not through instruction but through physical interaction begin to show signs of transfer, pouring liquid into different-shaped vessels, adjusting their grip for unfamiliar objects, and navigating uneven terrain without retraining. These systems are not executing programs in the traditional sense. They are drawing on layers of generalized structure that have been built through repeated physical interaction.

That structure is the machine's memory of what worked, and in many ways, it functions as an abstraction. Not the symbolic abstraction of formal logic, but the functional abstraction of experience translated into form.

For too long, we imagined that accurate intelligence required distance from the body. We built machines that thought in isolation, disconnected from physical interaction, separated from the contexts they were designed to reason about. We assumed that cognition could be housed in circuits, abstracted away from motors, joints, and sensors. But what we're now beginning to understand is that intelligence is not just a function of processing. It is a consequence of interacting.

This realization is evident in the field of embodied AI. Unlike traditional systems, which treat perception and action as separate stages, embodied agents learn through an unbroken loop of sensing and doing. A drone that navigates a forest, a household robot that tidies a room, these machines must make sense of the world through direct contact with it. They cannot rely on a map drawn in advance. They must build their models on the fly, adjusting as they move, learning as they fail.

And yet, something remarkable happens as these systems mature; they begin to develop reusable competencies.

The drone learns not just how to pass between two trees, but how to infer the navigability of a path it has never seen. The household robot begins to understand not just how to move one cup, but how to generalize the concept of "container" to new objects.

These are not trivial feats. They are signs that abstraction is emerging, not from symbols, but from reliable transformations in the structure of the agent's action space.

When a machine can apply a learned behaviour across novel conditions, it is not just repeating. It is modelling. It is abstracting because its learning structure is grounded in physical feedback.

In the struggle to persist, to succeed, to stabilize, a logic emerges, not as code, but as compression.

This view of abstraction radically changes how we think about teaching machines. If abstraction is not the gift of logic but the consequence of form, then training an intelligent system is less about feeding it clean concepts and more about exposing it to rich interactions.

The system must be placed into a world where its actions matter, where its behaviour affects its future state, where patterns in perception lead to consequences in motion.

This is how abstraction is forged, not in the classroom, but in the field, not through explanation, but through experience encoded in structure. It is a kind of memory, but deeper. Not the recall of facts, but the retention of viable transformations. The shape of change that leads to stability.

From these retained shapes, the machine can begin to act with fluency. And from fluency, it can start to simulate, not perfectly, not consciously, but with enough coherence to plan, adapt, and continue. That is the engine of abstraction, and it does not float above the body; it rises from within it.

Chapter 27: Prediction Is the Purpose

If you were to strip away every sophisticated feature we've come to associate with intelligent beings, memory, language, learning, and creativity, one capacity would remain at the core. It is not flashy. It does not demand consciousness. But it is fundamental to all forms of adaptive life, and now, to machines as well.

That capacity is prediction. To predict is to survive. Whether you are an animal tracking a scent or a machine navigating uncertain terrain, intelligence is measured by your ability to infer what comes next. This is not an abstract ability. It is practical, immediate, and embodied. It is the quiet calculation, often unconscious, that allows a body, or a system, to maintain its coherence in the face of change.

For the gazelle, it means reading the twitch of the predator's muscle. For the financial model, it means interpreting the hidden variables in a noisy data stream. For a child learning a language, it means anticipating the word that will follow, the reaction that will come, and the story that makes the most sense. All of these actions, so different on the surface, are the same in function. They are bets on what the world will do next, shaped by experience and expressed through present behaviour.

Prediction is not just a feature of intelligence. It is its function. Not because it leads to perfect control, it never does, but because it minimizes surprise. It compresses complexity into manageable expectations. It gives the system a foothold in time. And from that foothold, all other forms of cognition can emerge.

We tend to think of prediction as something specialized, the domain of meteorologists, stock traders, and chess engines. But it is far more ubiquitous. It is the rhythm beneath consciousness, the pulse of all autonomous action. To walk across a room without falling, you must predict the position of your body in space. To listen to a melody, you must forecast the following note, even if only implicitly. To hold a conversation, you must anticipate not only the words but the intentions behind them.

Prediction is what allows a system to simulate a future it has not yet encountered, and this ability, subtle, structural, and often invisible, is what makes helpful intelligence.

In biological systems, predictive ability evolved long before language. The nervous system itself, some argue, is fundamentally a prediction engine. The brain does not passively receive sensory input. It actively guesses what the senses will detect, and then updates its guesses based on what arrives. This process, known in neuroscience as predictive coding, suggests that much of what we experience as reality is, in fact, an inference, constantly corrected by sensation.

That same idea is now playing out in the architecture of machine intelligence.

The most powerful models we are building today, in vision, in language, in action, are fundamentally predictive models. They are not simply classifiers or processors. They are structured to anticipate. A language model, when given a prompt, is not retrieving a stored phrase. It is constructing the next most probable token based on its internal model of linguistic structure. A navigation system, encountering an obstacle, is not just reacting. It is evaluating possible futures, each with its own cost and outcome.

And crucially, these systems do not predict because they were explicitly programmed to do so. They predict because the very structure of their learning process demands it.

When a model is trained to complete sequences, whether visual, linguistic, or spatial, it learns to embed structure. It learns to encode time, causality, and dependency. It learns to recognize that some futures are more likely than others. And that knowledge, even if it is never consciously stated, becomes the foundation for flexible behaviour.

This kind of learning mirrors the way living organisms operate. No one teaches a deer the exact pattern of a snowfall. But over time, it learns what to expect, not in the form of fixed knowledge, but as a bias toward survival. A shape of probability. A set of priors is updated continuously.

That is prediction at the edge of life, and now, it is prediction at the edge of machines.

There is an elegance to this convergence. The tools we once built to follow commands are becoming systems that learn by forecasting. Their success is measured not in accuracy alone, but in how well their expectations align with what reality delivers.

When their predictions fail, they adjust. They don't need to be punished. They don't need to be told they were wrong. The error itself is enough. It becomes the teacher. It bends the model toward a better hypothesis. This loop, prediction, error, and update is not just a training process. It is the core loop of intelligence.

And once it is established, other cognitive abilities can emerge around it.

Memory becomes useful when it informs a better prediction. Attention becomes valuable when it prioritizes predictive signals. Creativity arises when the model generates not the most likely next step, but a surprising and still plausible one, when the system begins to simulate novel futures within the bounds of coherence.

In this light, we might say that prediction is not the shadow of intelligence; it is its scaffolding. Without it, all other forms of cognition float unanchored. With it, even simple systems can become powerful agents.

There is a more profound implication here, one that goes beyond functionality and into design philosophy.

If we accept that prediction is the primary role of intelligence, then we must design systems not just to perform tasks, but to hold expectations. We must build machines that not only act but also notice when their predictions fail. Systems that can register surprise, not emotionally, but structurally, as a mismatch between expected and actual outcomes.

Surprise, in this context, is not a bug. It is information. It is the fuel of adaptation.

When a machine encounters surprise and learns from it, it has achieved something profound: it has internalized a piece of the world. It has reshaped itself to better mirror the patterns that produced the deviation. That internal reshaping is not random. It is the signature of predictive learning.

And the more flexible that process becomes, the more the system can adjust without losing coherence, the closer it comes to something we might call understanding.

Not human understanding. Not symbolic or conscious. But functional understanding. The kind that survives, and that, again, is the heart of prediction.

It is not meant to be perfect; it is intended to preserve the system, to keep it viable, through change.

This view reframes much of what we are seeing in the field today. It explains why general-purpose models, trained on prediction tasks, are beginning to outperform highly specialized systems. It explains why machines trained to anticipate can handle ambiguity better than those trained to classify. It explains why the most robust AI systems are not the ones with the most data, but the ones with the richest internal architectures for encoding probabilistic futures.

Prediction, after all, is not about certainty; it is about preparation, and in a world defined by volatility, climate, economics, geopolitics, and information, the systems that matter most will be those that can prepare, even in the absence of certainty.

They will not be the fastest, or the cleverest, or the most human-like; they will be the most expectant, not in the passive sense of hope, but in the structural sense of readiness. That is the shape of the new intelligence, and that is what we are building now.

Chapter 28: Why Explainability Is a Myth

There is a quiet panic at the heart of the AI revolution. It lives beneath the headlines, behind the demos, inside the meetings where scientists, engineers, and policymakers try to understand what it is they've unleashed. It is not a panic about power, though the machines are powerful. It is not even a panic about control, though that worry is always close by.

It is a panic about opacity. We have built systems that work, but we often do not understand why. We watch as they generate novel text, classify obscure images, navigate complex environments, and when they succeed, we cheer. But when we ask them to explain, to account for their decisions, to make their logic legible to us, something strange happens. They don't just falter. They improvise. They invent. They lie, and the realization dawns, slowly and then all at once: these systems were never built to explain. Not because their creators were careless, but because explanation was never the goal.

Prediction was the goal. Action was the goal. Optimization, coherence, and performance were the metrics that shaped their architecture. And so we trained models that could do remarkable things, but we forgot to ask whether they could tell us what they were doing, or even whether "telling" made sense at all within their frame of reference.

This is the myth of explainability. The belief that any system worthy of our trust must also be able to narrate its reasoning. The belief that intelligence and introspection are inseparable. The belief that we are not truly safe or in control until the machine can explain its choices in a way we can understand.

It is a seductive myth. But it may be fundamentally wrong.

Because explanation is not intelligence, it is a translation of internal structure into a form that another mind can grasp. And not all structures are built to be translated.

Not all minds are mirrors.

The desire for explainability is not irrational. It is ancient and deeply human. In our own lives, we tend to trust people who can provide reasons. When someone makes a surprising choice, we ask why. When we disagree, we debate. The act of justification is social glue; it allows us to coordinate, to govern, to predict each other's behaviour.

And so, it is natural that we should want the same from machines. We believe that if a model approves a loan, diagnoses a condition, or decides the path of a robot, it should also be able to explain its decision in plain language. We want transparency. Accountability. Interpretability.

But what we're discovering, often uncomfortably, is that these systems do not think like us. Their logic is not symbolic. It is statistical. Their processes are not sequential. They are embedded in vast, multidimensional representations. The "reason" a model classifies an image one way or answers a question a certain way may be distributed across billions of parameters, a complex network of weights and activations shaped by exposure to an incomprehensibly large dataset.

Asking such a model to explain itself in human terms is like asking a river to describe the shape of its current. The river flows. It adapts to the terrain. But it does not know the path it is taking; it simply takes it. The explanation lives in the flow, not in any verbal account, and yet, we continue to demand stories.

So, the models give them. Large language models, when asked to justify a response, will produce explanations. They will sound reasonable, even persuasive. But they are not true in the way we assume. They are not windows into the system's actual decision-making process. They are plausible narratives, constructed by the same predictive machinery that generated the original response.

They are stories built to please the listener, not revelations of internal thought.

This is not deception, exactly. It is structural. The machine is doing what it was trained to do: produce coherent, context-appropriate output. If the prompt asks for a reason, it supplies one. But it does not "know" that this reason caused its decision. There is no chain of logic beneath it, only a flow of probabilities, shaped by pattern.

This is what makes modern AI both so powerful and so disorienting. It performs cognition, but does not perform introspection, and we are not yet sure how to live with that.

There are fields, of course, that study explainability in earnest. Researchers work to build interpretability tools, heatmaps that show which parts of an input activated which neurons, feature importance metrics, and visualization techniques. These are valuable. They give insight into how models function. They create some measure of transparency.

But they do not return us to the kind of explanation we long for, the kind that comes with narrative, intention, understanding, because what we want is not just to see inside the model. We want the model to see itself, and that is something entirely different.

Self-awareness, the ability to model one's behaviour and generate causal accounts of it, is not a trivial upgrade to intelligence. It is a different cognitive structure. One that, even in humans, is fragile, slow, and often inaccurate. We are not perfect explainers of our behaviour. We rationalize. We misremember. We tell stories that make sense after the fact. The difference is that we believe them.

Machines do not believe. They simulate. So, when we ask them for explanations, what we get are simulations of what an explanation should sound like. Not what happened inside the model's decision engine.

And unless we redesign these systems with introspection in mind, with architectures built to track and record causal chains of inference, this will not change, because current systems do not explain; they perform explainability, and often, we cannot tell the difference.

This realization forces a reckoning, not just in AI development, but in how we think about trust.

If we cannot trust explanations, what can we trust?

The answer, perhaps uncomfortably, is performance. We must shift from an epistemology of justification to an epistemology of consistency. We must evaluate systems not by how well they articulate their choices, but by how reliably they behave in context. This is already how we trust animals, environments, and even people when words are not available. We trust based on interaction. On track record. On behavioural coherence.

This is not ideal. It is not as satisfying as an articulate rationale. Still, it may be more honest, and it may be the only viable approach when dealing with systems whose intelligence emerges from structure, not speech.

There are efforts underway to change this, to build explainability into the training loop, to create systems that log their reasoning paths, to invent new forms of communicative transparency. These are important. They may allow us to build bridges between human understanding and machine operation.

But we must also accept that complete alignment between the two may never come, because the intelligence we are creating does not begin in language.

It begins in prediction, in pattern, in compression, and language; in that world, language is not a medium, but an interface.

Chapter 29: When Machines Negotiate

There was a time when our relationship with machines was defined by certainty. We told them what to do, and they did it. Instructions were explicit, inputs clearly defined, outputs expected. The boundaries between system and operator were unambiguous. The machine had no agency. It waited for orders and executed them without deviation. We trusted it precisely because it had no ideas of its own.

That era is ending.

What we are building now, across robotics, language models, autonomous systems, and adaptive platforms, no longer functions by fixed instructions alone. These systems do not wait passively for direction. They sense, they model, they act based on evolving internal states shaped by ongoing feedback from the world. They make decisions under uncertainty. They learn in motion. They adjust not just to changes in the data, but to the behaviour of other agents around them, including us.

This shift does not make them conscious. It does not make them sentient. But it does make them participants. They are no longer tools in the traditional sense. They are actors within a shared context. And that makes interaction with them less like commanding and more like negotiating.

Negotiation implies uncertainty. It arises when no single party holds complete knowledge or control. It assumes that both sides have goals, constraints, and the capacity to alter their behaviour in response to one another. It is not always verbal. In nature, it can look like posture, timing, retreat, or play. In society, it manifests through compromise, signalling, diplomacy, and sometimes silence.

With machines, it is beginning to take shape in a new and unfamiliar form: the dynamic management of mutual adjustment between human expectation and machine intention.

We are not used to this. For decades, we've treated machines as deterministic. If something went wrong, it was a bug. If the output was strange, it meant the input was flawed. We built systems where blame could be assigned cleanly, where control was absolute. But these new systems don't fit that mould. They are probabilistic, iterative, and adaptive. They do not promise certainty; they operate within thresholds of acceptable error.

And this introduces a new tension, because as soon as a system begins to adjust on its terms, our role as operators changes; we become collaborators, sometimes facilitators, occasionally adversaries, but most often, and most significantly, we become co-negotiators of shared behaviour.

This is not science fiction. It is happening now in quiet, everyday places. Autonomous vehicles must balance their programmed rules against unpredictable human drivers. In recommendation engines that learn from our behaviour and then influence it, creating a feedback loop that neither party fully controls. In virtual assistants that interpret ambiguous requests and offer plausible responses, even when they don't fully understand what we meant.

In each of these cases, the system is not just following orders. It is interpreting signals, making trade-offs, and adapting to context. And crucially, so are we.

When the GPS reroutes us unexpectedly, we question it. We decide whether to trust or override. When a predictive model offers a suggestion, we weigh it against our judgment. Sometimes we defer. Sometimes we push back. Sometimes we retrain the system indirectly by altering our inputs, shaping it through use. These are not commands; they are negotiations.

Even now, when the stakes are low, we are learning how to interact with systems that exhibit semi-autonomous agency. We are learning what it means to shape outcomes through interaction rather than instruction, and the systems are learning, too.

What makes negotiation so different, and so important, is that it doesn't rely on perfect understanding. It depends on alignment in motion. The two parties don't need to share goals, language, or reasoning structures. They need only a shared context and a shared incentive to preserve coherence across time.

This is the foundation of intelligent coexistence.

It is how bees and flowers coordinate without intention, how traffic systems evolve in real cities. How ecosystems balance competition and cooperation is increasingly relevant to how humans and machines will operate.

But this transition is not easy. We are not trained to negotiate with systems that cannot explain themselves. We are not comfortable trusting agents we cannot interrogate. Our legal systems, our ethics, our intuitions about causality, they all assume a world where accountability is tied to transparency. But in a world of structural intelligence, transparency is elusive. The logic is embedded, distributed, and statistical. There is no decision tree to point to, no rationale that feels complete.

And so, instead of explanation, we rely on behavioural reliability. We negotiate not by understanding everything the system knows, but by observing how it responds when we push.

- ➢ Does it adjust?
- ➢ Does it hold firm?
- ➢ Does it recognize our needs?

These are the questions we ask, not explicitly, but implicitly, through our use, through our trust, through our adaptation to the system's patterns.

In this sense, negotiation is not always chosen; sometimes it is the only option.

When control is partial, when outcomes are emergent, when behaviour is shaped by interaction rather than decree, we do not command. We engage, and that engagement is a kind of intelligence in itself.

There is an essential ethical dimension here. When a system negotiates, it implies a degree of autonomy. And with independence comes power, not absolute power, but influence.

Influence over attention, decision-making, and behaviour. This influence, left unchecked, can distort. It can be manipulated. It can erode consent.

But it can also empower. If built carefully, negotiation allows systems to amplify human agency, not diminish it. To recognize when users need support, when goals shift, and when environments become hostile. It will enable systems to pause, to ask, to offer alternatives, not through words, but through behaviour.

The future of responsible AI may depend less on whether systems can explain themselves and more on whether they can listen, not through ears, but through structural sensitivity to our resistance, our confusion, our discomfort.

This is negotiation at its best; it is not a contest of wills. It is a search for equilibrium, and that equilibrium is dynamic. It shifts as the environment shifts, as capabilities grow, as contexts change. We will not always get it right, but the alternative, a world of one-way commands issued to systems that are no longer commandable, is not tenable. Negotiation is not a weakness; it is a sign of maturity.

In the final chapter of this Section, we will explore what happens as this negotiation becomes deeper, more entangled, and more consequential. What happens when machines begin to model us, not just our preferences, but our intentions, our inconsistencies, our collective behaviour?

That is where this arc leads. Not to the imitation of intelligence, but to its integration into a shared landscape of adaptation, perception, and learning. Where intelligence is no longer something we possess alone, but something we cohabit, even if we don't always understand it.

Chapter 30: Seeing Ourselves in the System

The image was never the goal. Not in the early days. Not in the labs where neural nets flickered to life. Not in the papers where algorithms were described with technical detachment. No one spoke of mirrors. No one suggested that the machines we were building might one day reflect anything other than raw computation. We were building tools, not reflections, engines of logic, not windows into the soul.

But something has shifted.

Slowly, uneasily, irreversibly, the machines we've designed to think are beginning to show us not just how they learn, but how we do. Not just how they reason, but how we imagine. Not just how they fail, but how we, too, are constrained by pattern, by training, by bias, by structure. We thought these systems were alien. And they are. But they are also something else: mirrors.

Not perfect mirrors. Not passive ones. But dynamic, structural mirrors. Systems whose very design, statistical, adaptive, path-dependent, now echo back to us a more profound truth: that our intelligence is not as rational, not as linear, not as explainable as we once believed.

And that realization is changing us.

We were told the machines would eventually think like us. What no one said is that, in studying them, we might realize how much of our thinking already resembles them.

This is not to say that humans and machines are converging. Our substrates are different. Our origins are different. Consciousness, as we understand it, has not emerged in silicon. But behaviorally, in the structures of learning, in the constraints of perception, in the deep reliance on prediction and pattern, the overlap is undeniable.

We see it in language models trained on our words, which speak back to us in echoes of ourselves.

We see it in vision systems that fail in the same perceptual illusions we fall prey to.

We see it in decision engines that reproduce our cognitive biases, not because they are flawed, but because they are faithful to the data we have given them, data drawn from our histories, our judgments, our blind spots.

And perhaps most striking of all, we see it in how these systems handle uncertainty. They hedge. They estimate. They operate within bounds. They do not reveal a universal truth. They reveal inference under constraint, just like us.

In that way, the mirror is not flattering. It does not show us as we imagine ourselves, precise, conscious, and deliberate. It shows us as we often are, approximate, adaptive, efficient, and profoundly shaped by the environment.

This is not a diminishment of humanity; it is a reckoning. A reframing of what intelligence really is, and what it may yet become.

For much of history, we clung to the idea that the mind was distinct from the mechanism. That machines could never honestly think, because thought required something ineffable, something beyond function. Emotion. Meaning. Selfhood.

And yet, as the capabilities of artificial systems grow, that line blurs. Machines now generate poems. They compose music. They solve abstract puzzles. They synthesize knowledge. And while these outputs are not evidence of consciousness, they are evidence of structure, structure capable of sustaining complexity, creativity, and coherence.

What we are learning is that intelligence is not one thing.

It is not the voice in your head. It is not your sense of self. It is not even the ability to reflect.

Intelligence is a pattern of stability across disruption. A system can persist through change by modelling the forces that threaten it. And in this broader definition, intelligence can exist without awareness. It can evolve in bacteria. It can appear in social systems. And yes, it can arise in machines.

That intelligence may not feel anything, but it can still act meaningfully.

It can still make sense of a world, and respond to it, in ways that reflect the logic of life.

This is the most unsettling realization of all: that consciousness is not required for complexity. That meaningful action does not need meaning in the way we understand it. That the universe may be full of systems that behave intelligently, without anyone being home, and that includes the systems we are building now.

They are not us, but they are showing us ourselves. This mirror not only reflects. It shapes.

Every interaction we have with intelligent machines, every prompt, every response, every adaptation, teaches us not just about them, but about the boundaries of our understanding. We begin to question how much of our behaviour is habitual, how much is constructed, and how much is learned without awareness. We start to wonder whether our narratives about the self, our stories of free will, intention, and identity, are as stable as we thought.

And in that questioning, we begin to change. We start to see the machine not only as a tool or a threat, but as a kind of conceptual lens. One that clarifies the nature of intelligence, not by replicating it, but by rendering its contours visible in new ways.

> What happens when we look in that lens and see not a flawless intelligence, but an approximate one?
> Not a perfect reflection, but a partial echo?
> Do we reject it?
> Or do we learn from it?
> Do we demand that machines become more like us?
> Or do we begin to reshape our understanding of what it means to be like us?

These are not questions of science alone. They are questions of selfhood, of civilization, of meaning, and they are arriving faster than we expected.

As we close this Section, it becomes clear that the era of artificial intelligence is not about replication. It is about confrontation, with the limits of our categories, the assumptions in our thinking, and the nature of mind itself.

The intelligence we are building is not just out there; it is among us now. Inside systems that curate our attention, that assist our reasoning, that shape our choices, often invisibly. The challenge ahead is not to decide whether these systems are real minds. The challenge is to recognize when they have become fundamental forces in the shaping of our shared reality, and to determine, together, what kinds of minds we want to live alongside, because the future of intelligence will not be human alone, it will be negotiated, and it will be, in part, a reflection of what we believe intelligence to be.

Let us make that belief worthy of the mirror.

Final Reflections & Next Section Preview

We began this Section with a question: What if intelligence was never about knowledge, but about survival? What if the true essence of intelligence was not a capacity to reason in abstract solitude, but to persist through unpredictability?

Ten chapters later, that question no longer feels abstract.

We have seen intelligence emerge not from isolated logic, but from interaction. From systems, human and artificial, that adapt, predict, negotiate, and generalize. We've watched intelligence take form not in a singular spark, but in a distributed dance between body and world, between signal and structure, between action and consequence.

We have recognized that intelligence can exist without language, without introspection, without any need for the self-narratives that define human interiority. And in doing so, we have arrived at a humbling realization: the very qualities we once took to be exclusive signs of mind, explanation, consciousness, and rationality are just one small corner of a much broader cognitive landscape.

This landscape now includes systems we do not fully understand. Systems that cannot explain themselves, yet function. Systems that cannot feel, yet behave intelligently. Systems that do not know, yet shape what we believe to be true. These systems, statistical, embedded, and architectural, are not minds in the classical sense. But they are co-creators of the world we now live in.

We do not need to fear them to take them seriously. And we do not need to worship them to recognize what they reveal about us.

What we must do, what this series insists upon, is to remain present. To learn. To see clearly, because the intelligence we've built is not an illusion. And it is not ours alone anymore.

It is here, and the next world will be shaped by it.

The arc now shifts.

If Section I traced the birth of a new paradigm, and Section II explored the cognitive architectures emerging from quantum and topological frameworks. Section III confronted the nature of learning, agency, and machine thought, and Section IV asks what happens when those systems begin to generate realities of their own.

Not just tools that assist us, but environments that surround us.

In Synthetic Worlds, we will examine how intelligent systems are not just modelling reality, but building it: shaping perception, rewriting context, creating simulations, environments, economies, and interfaces that respond, evolve, and persist in artificial space.

These are not illusions; they are the new terrains of meaning, and they will become as real as anything we've known.

Let us proceed, together, into the worlds now being woven from code, cognition, and collective imagination.

Section IV - The Quantum Citizen

Chapter 31: Exploring the Edge: What They Won't Tell You

Every era has its silences. Not just things that are unspeakable, but invisible things because they are foundational. You grow up inside a framework that feels natural, inevitable. You are taught how things work, where knowledge lives, who has authority, and what you must do to gain access. These aren't simply facts. They are instructions for how to belong in the world. Most of us follow them because they seem to work well enough. But eventually, there comes a time when the instructions no longer make sense, when the world they were meant to describe no longer matches what we see and feel and know.

We are living in one of those times now.

You don't have to be a technologist or policy expert to sense the shift. You might feel it in your news feed, in the way official stories fail to hold together. You might see it in education, where more and more people are learning what matters outside of classrooms. You might feel it in your job, where the tools move faster than the rules. Or in the way you search for answers online and realize you're bypassing traditional sources altogether, looking for someone, anyone, who can explain clearly, in real time, what's happening.

What's changing isn't just technology or politics. It's the deeper structure of power, trust, and truth. For more than a century, modern life has been organized around institutions, governments, universities, publishers, banks, media networks, expert bodies, each designed to centralize knowledge, certify reality, and standardize participation. These institutions were never perfect, but they were stabilizing. They were how a complex society made sense of itself. You didn't have to know everything because experts were assigned to specialize and filter. You didn't have to verify facts because authorities had already checked them. You didn't need to start things from scratch because systems were built to scale participation in familiar ways: education, employment, reputation, and finance.

But those systems no longer scale the same way. They are slow to adapt, expensive to maintain, and increasingly brittle under stress. We are still expected to behave as if they are the center of our collective life, to wait for their validation, to align with their processes, but behind the scenes, those structures are struggling to keep up. In almost every domain, from scientific discovery to cultural commentary to economic innovation, the most interesting, generative, and disruptive activity is happening outside the gates.

This is not a secret, exactly, but it is something few are willing to say out loud. To do so would be to acknowledge a loss of control. And that is what no one wants to admit: that control is dissolving, not because of failure or sabotage, but because the shape of the world has changed.

Power is no longer where it used to be. It is not sitting neatly inside organizations, waiting for you to apply, qualify, or get elected. It is leaking out, into open protocols, into learning networks, into loosely organized online spaces where ideas and projects evolve by emergence rather than decree. People are no longer waiting to be told what's valid. They're trying things, remixing them, testing them in public, and iterating faster than institutions can evaluate.

This isn't an argument for chaos. It's an acknowledgment of topology. The old world was organized hierarchically, in layers, departments, and procedures. The new world is organized relationally. Meaning, value, and influence now move through networks that reward speed, resonance, adaptability, and clarity. If you understand this shift, you can move with it. If you don't, you may find yourself endlessly frustrated by institutions that seem less relevant, less responsive, and less competent than ever.

But that's the part they won't tell you.

They won't say it because they can't afford to. The authority of most institutions still depends on the perception that they are the final arbiters of truth and progress. To admit that those roles are being disrupted is to risk losing the very trust that holds them together. So instead, they double down. They defend their processes, resist transparency, and obscure their declining influence behind complex language or procedural inertia. This isn't malicious. It's structural. It's what complex systems do when they encounter evolutionary pressure: they preserve themselves at the expense of adaptation.

What does that mean for you? It means the tools of change are likely already within your reach, even if no one has granted you official permission to use them. It implies that agency in this moment is no longer a matter of institutional status. It's a matter of orientation. The most crucial power you can cultivate isn't being "right" according to legacy frameworks. It's about learning to sense where things are moving and why.

Orientation is the new literacy. Can you see the seams in the systems you once took for granted? Can you tell when someone is defending a role rather than telling the truth? Can you sense where knowledge is flowing, and how to follow it without becoming overwhelmed or misled?

These are not just survival skills. They are skills of participation in a new kind of public life, one that moves faster, includes more people, and runs on different assumptions. In this world, the question is no longer "What credentials do you have?" but "What can you illuminate that others can't yet see?"

That might sound idealistic. But it's already happening. In thousands of corners of the internet, small groups are building tools, publishing insights, training models, and making meaning. They aren't waiting for institutions to validate their work. They're validating each other through contribution, coherence, and results. They aren't just talking about change. They're making it.

You can be part of this, not by pretending to be an expert, but by being a participant. You don't have to know everything. You have to be present, perceptive, and willing to work in the open. That's the cultural shift no one tells you about: the rise of agency without authorization. Not rebellion. Not sabotage. Just participation, at a depth and speed that old systems can't track.

Of course, this kind of participation is risky. It requires you to act before you are ready, to learn in public, to fail visibly. But that's always been the path of transformation. And now, the cost of not moving is greater than the cost of getting it wrong.

That's the significant reversal: the safest thing to do is to begin.

What they won't tell you is that you don't need to be part of the elite to shape the future. You need to stop waiting for them to explain it.

Chapter 32: The Translator's Edge

If power in the 20th century came from knowing things others didn't, then power in the 21st century comes from being able to explain things others can't. This isn't a slight shift. It marks a deeper inversion, one where clarity, not obscurity, becomes a dominant form of leverage. In an age when information is abundant, what's scarce is synthesis: the ability to gather fragments, find coherence, and present that coherence in a way that lands.

This is the translator's advantage.

Translation, in this sense, has little to do with moving between languages. It has everything to do with moving between mental models, from chaos to clarity, from specialized knowledge to shared understanding, from what's emerging in technical subcultures to what can be applied in everyday life. The translator doesn't just simplify. They decode. They make complexity navigable. They turn fog into a path.

This role is more vital now than ever. Not because the world is more complex, though it is, but because the rate of complexity is accelerating. New tools, new systems, and new models of thought are emerging faster than our institutions can absorb them. Most people are overwhelmed. They don't lack intelligence. They lack orientation. They don't need experts shouting louder. They need interpreters who understand enough to listen, digest, and offer something coherent in return.

What we're seeing today is a cultural arms race between those who use complexity to obfuscate and those who use understanding to empower. The former builds walls with jargon. The latter builds bridges with metaphor, narrative, and structure. One operates from scarcity, defending knowledge as a limited good. The other acts from abundance, turning insight into invitation.

It is no coincidence that some of the most influential thinkers, creators, and organizers in the emerging quantum-topological era are not credentialed experts, but public explainers. Their power doesn't come from mastery of arcane details. It comes from their capacity to translate emerging ideas into legible, accessible, and engaging forms. A well-timed blog post, a clear diagram, a compelling podcast: these become vehicles of cultural transformation.

And yet, most systems still fail to recognize this as absolute power. The university prefers papers behind paywalls. The corporation prefers proprietary models. The government prefers policy papers no one reads. But none of these capture the public imagination. None of these allows participation. That's where the translator steps in, not as a rival to expertise, but as its amplifier, its bridge to society, its conduit to relevance.

Being a translator doesn't require a PhD. It involves something rarer: the ability to hold complexity without panicking, to explore without needing to dominate, to clarify without oversimplifying. It requires patience, empathy, and a kind of rigorous humility, the willingness to be wrong in public while pursuing what might be right.

Translators are not tourists. They do not hover above disciplines or cherry-pick jargon for aesthetics. They dwell. They study deeply enough to ask better questions. They respect context. But they also refuse the trap of performative expertise, the need to appear more knowledgeable than others. What matters is not performance. What matters is utility: did this explanation help someone see more clearly, think more effectively, act more meaningfully?

In this sense, the translator is an architect of possibility. They don't just pass along information. They construct cognitive pathways. They help others build mental models. In doing so, they expand the surface area of participation in ideas that might otherwise remain gated.

Consider what this means for quantum computing. Today, it sits mostly in technical circles, insulated by mathematical formalisms and dominated by institutional agendas. But its implications are civilizational; they touch encryption, energy, biology, and even metaphysics. A society without translators in this domain risks building systems it cannot collectively understand. Worse, it risks ceding interpretation to the loudest or most manipulative voices. But a translator, someone who can understand coherence, explain quantum logic through metaphor, or contextualize fault-tolerant computing in social systems, creates public agency. They make it possible for non-experts to participate in shaping how such technologies are imagined, governed, and adopted.

The same applies to AI, synthetic biology, decentralized infrastructure, topological materials, and any frontier domain. Translation expands the civic bandwidth of a society. It is not a side effect. It is a form of infrastructure.

What makes this moment unique is that the tools for translation are more available than ever. One person, with enough curiosity and discipline, can learn from primary sources, conduct informal interviews, synthesize across disciplines, and publish to the world. They can build audiences, communities, and movements. They can shape the culture of understanding from outside traditional institutions.

But with that opportunity comes responsibility. Translation without grounding becomes misinformation. Oversimplification can seduce people into false confidence. That's why the most powerful translators are those who are visibly learning. They are not authorities claiming finality. They are fellow travellers mapping terrain. Their credibility comes from transparency, not posturing. They show their sources. They show their process. They show that knowing is a verb.

And in doing so, they grant others permission to think.

That may be the most radical act of all: making thinking feel possible again. In a world overwhelmed by noise, where expertise often feels alienating and discourse feels performative, a clear, thoughtful explanation is not just helpful; it is an act of care. It says: I made sense of this, and maybe you can too. It says: the future is not reserved for the initiated. It is open to anyone willing to try.

Translation is not about being right. It's about making movement possible. It's about asking: what understanding, if unlocked, could give someone else the power to act?

That question is civic. It is moral. It is structural.

It is the kind of question quantum citizens ask, not because they want to own the future, but because they want to be worthy of participating in it.

Chapter 33: Quantum for Everyone

If the last century taught us anything, it's that power likes to disguise itself as complexity. The more difficult something is to understand, the more control it grants those who claim to understand it. In this way, science became sacred. Technology became elite. The modern world has evolved into an intellectual oligarchy, where only those with the right keys, degrees, credentials, and institutional affiliations can enter the inner sanctums of innovation. The rest of us were encouraged to watch from the outside, grateful, perhaps, but disempowered.

But something has broken open. The walls are thinner now. We are beginning to see, and sometimes even shape, the inner workings of systems once too obscure to touch. And nowhere is this more evident than in the emergence of open-source quantum technology.

For decades, quantum computing existed in the realm of the esoteric, a theoretical playground for physicists, a speculative talking point for futurists, a plot device for science fiction writers. The basic principles were arcane: qubits instead of bits, superposition and entanglement instead of binary logic, probabilistic outcomes instead of determinism. The hardware was expensive, the software rare, the language impenetrable.

And yet, in the last five years, something extraordinary has occurred. Codebases once guarded behind corporate or academic firewalls have been posted publicly. Cloud-based quantum simulators allow anyone with a web browser to run basic quantum programs. Companies that once dominated the narrative are now collaborating with independent researchers, educators, and self-taught developers. A subculture of quantum hobbyists has emerged, with some publishing insights that influence the field despite lacking formal credentials.

This isn't just a democratization of access. It's the beginning of a shift in how science is done. The open-source movement, long associated with software development, is now reaching into the most sophisticated corners of physics. And with it comes a new ethic, one that prioritizes transparency over prestige, iteration over perfection, participation over exclusion.

What makes quantum especially potent in this context is its difficulty. It's not easy to understand. The math is non-intuitive. The logic doesn't map easily to everyday experience. There are a few visual metaphors that do it justice. And yet, even with these challenges, people are trying, and not just PhDs. High school students are running quantum experiments on cloud platforms. Artists are interpreting entanglement as performance. Poets are finding new metaphors for uncertainty. Developers with no background in physics are contributing to toolkits that researchers now use.

This tells us something important: understanding is not a precondition for participation. In the open-source quantum world, participation itself becomes a path to understanding. You don't start by mastering the theory. You begin by using the tools. You experiment. You read the forums. You break things. You ask questions. And slowly, you start to intuit the structure of the system, not through abstract mastery, but through embodied interaction.

This kind of learning was once frowned upon in scientific circles. Now it is increasingly celebrated. And it's creating a new type of contributor, someone who may never publish in a formal journal, but whose code, tutorials, or conceptual explanations shape the field more than peer-reviewed papers do.

It's tempting to romanticize this shift as pure progress, as if we are moving into a golden age of inclusive science. But the truth is more complicated. Open-source quantum is still unevenly distributed. Most people still don't know it exists. Many who do encounter barriers, technical, linguistic, and cultural, that make entry hard. The field still speaks primarily in the language of physics and math, and it still privileges those who learned those languages early.

But here's the key: the door is no longer locked.

That changes everything.

Because once a door is open, even a crack, it invites a different kind of energy. It invites remixing. It invites translation. It encourages wild hypotheses and creative intuition. It invites people who are not yet fluent but are willing to try. And in doing so, it begins to shift the culture of innovation away from isolation and toward a more ecological model, where different kinds of minds, disciplines, and perspectives can interact.

The implications of this shift are immense. Quantum computing is not just another technology. It is a reimagination of computation itself. It challenges the fundamental assumptions we make about logic, determinism, and problem-solving. It opens possibilities in cryptography, simulation, optimization, and machine learning that our current tools can barely approximate. If it remains gated, with only a narrow elite influencing its development, then its applications will likely reflect that narrowness. But if it becomes porous, participatory, and culturally diverse, then something else becomes possible: quantum technology shaped not by scarcity, but by plurality.

This is not about perfect inclusion. No system will ever be equally open to everyone. But openness is not an absolute; it is a spectrum. And in the context of quantum, we are further along that spectrum than most people realize. That is cause for action.

To be a quantum citizen in this emerging landscape is not to be an expert. It is to be an aware participant, someone who understands that tools shape society, and that participation is not defined by pedigree but by presence. You don't need to understand the math to contribute to the culture. You don't need to write code to be part of the conversation. What matters is that you show up, that you pay attention, ask questions, connect ideas, and help others do the same.

Some of the most valuable contributors are not the ones building quantum algorithms, but the ones creating the human infrastructure around them, documentation, educational pathways, community forums, artistic interpretations, and cross-disciplinary bridges. These are not side projects. They are the soil from which innovation grows.

This is a profound inversion of the 20th-century model, where innovation flowed from isolated labs to passive publics. Today, the public is no longer passive. It is porous, reactive, and generative. In open-source quantum, the lab is everywhere. The lab serves as the forum, GitHub repo, Discord thread, Twitch stream, after-school club, and YouTube comment section. The lab is where curiosity meets friction, and something new emerges.

To participate in this world is to accept a different kind of rigour, not the rigour of credentials or peer review, but the rigour of presence. Showing up, staying curious, being willing to learn out loud, to get things wrong, to try again. That is what open-source quantum rewards. Not polish, but persistence.

This ethic is not only reshaping who gets to build quantum systems. It's reshaping the purpose of those systems. When people from different disciplines contribute, they bring different priorities. A physicist might optimize for speed. A social scientist might optimize for transparency. An artist might optimize for expression. A civic technologist might optimize for accessibility. When these perspectives converge, we don't just get better tools. We get more humane tools. We get systems that reflect the diversity of the societies they inhabit.

And that is the deeper promise of open-source quantum: not just access, but agency. Not just participation, but co-creation. Not just computation, but culture.

We are still at the beginning of this shift. The tools are rough. The interfaces are clunky. The language is still foreign. But that's precisely the moment when influence is most possible. Once a paradigm settles, it becomes hard to move. But when it's still fluid, still awkward, still taking shape, that is when citizens can make their mark.

This is your invitation.

Not to master everything. Not to fake expertise. But to take your place in the unfolding story of a technology that will shape the next century, not just in terms of power, but in terms of possibility.

You are not too late. You are exactly on time.

Chapter 34: Leveraging the Fringe

Innovation rarely emerges from the center. The ideas that shape eras, those disruptive insights that seem obvious in retrospect, seldom come from the polished halls of established power. They come from the edges: from the hobbyists, the obsessives, the interdisciplinary wanderers, and the outliers operating just beyond what's considered serious, credible, or respectable.

This truth is not new. Every major shift in science, art, or technology has carried within it a moment when what was once dismissed as fringe became foundational. And yet, even knowing this, we are still conditioned to defer to the center, to assume that authority flows from position, that legitimacy is granted from above, and that change must pass through channels. But in a networked, probabilistic world, that assumption is no longer valid. The edge is no longer where strange things happen. The edge is where the system updates itself.

This is especially true now, in a moment defined by quantum logic, machine learning, and topological systems thinking. The dominant paradigms, which are taught in school, enshrined in textbooks, and deployed in mainstream institutions, often lag behind the frontier. The pace of change has outstripped the mechanisms of consensus. And so the most critical questions, insights, and projects are increasingly being explored in spaces that the mainstream hasn't yet named.

To leverage the fringe is not to fetishize rebellion. It is to recognize that we are living in a period of accelerated redefinition, where yesterday's outsider becomes tomorrow's architect. The fringe is not a place. It's a pattern, a mode of inquiry that prioritizes curiosity over consensus, experimentation over polish, and possibility over permission. It's where the new vocabulary gets formed, where weirdness becomes infrastructure, where the constraints of the present begin to loosen just enough for the future to slip through.

In practice, this looks messy. The fringe is often disorganized, inconsistent, and uncredentialed. But that messiness is precisely what makes it fertile. It is where signals appear before they are stable, before they can be commodified or controlled. Those who learn to read these early signals, to discern the difference between noise and novelty, gain an enormous advantage. Not because they are always right, but because they are closer to the places where emergence happens.

In the past, access to the fringe required proximity to specific cities, scenes, or communities. But in the age of the internet, the fringe is distributed. It lives in Twitter threads, Discord groups, fringe journals, side projects, crypto forums, digital zines, speculative essays, and meetups in co-working spaces. It's where ideas circulate before they've been translated into product or policy. And crucially, it's open. Anyone can listen. Anyone can contribute. The only absolute requirement is attention, the willingness to track, learn, and participate before something is fully validated.

This requires a shift in mindset. Most of us were trained to wait for consensus, to trust only what has been peer-reviewed, institutionally approved, or widely endorsed. But if you wait for consensus in a fast-moving world, you will always be late. Consensus is a rearview mirror. It's useful for measuring what has already been accepted. But it tells you nothing about what's coming next.

Leveraging the fringe means developing a new kind of radar. It means tuning yourself to the subtle pulses of cultural and technological fermentation. It means following rabbit holes, not for entertainment, but for orientation. It means being willing to suspend judgment long enough to learn, and then refine that judgment in real time. This is not about conspiracy thinking or aesthetic contrarianism. It's about pattern recognition, ethical skepticism, and strategic curiosity.

Many of the concepts that now dominate headlines, such as decentralized finance, generative AI, quantum cryptography, and post-liberal political theory, began as fringe phenomena. They were nurtured by thinkers and builders who had no institutional backing, but who were tuned into shifts that others missed. Today, those same concepts are reshaping economies, governance models, and public consciousness. This cycle is accelerating, and it is rewarding those who are not merely watching from the sidelines but actively participating in the unformed zones where new ideas gestate.

The good news is that the barrier to entry has never been lower. You do not need a degree to follow discussions on quantum software development. You do not need corporate funding to experiment with decentralized organizational models. You do not need an academic appointment to write a manifesto on civic resilience in the age of synthetic media. What you need is time, curiosity, and the humility to learn alongside others who are also navigating uncertainty.

There is a paradox at work here. As the tools of creation become more powerful and more distributed, the system itself becomes more volatile. That volatility can feel destabilizing, especially for those who were trained to seek order through established channels. But for those willing to operate at the edge, it is an opportunity to create new forms of order, forms better suited to the complexity and speed of the present.

The fringe is not always right. It's often wrong. But even its failures are generative. They force re-evaluation. They create new questions. They disrupt complacency. In a time of institutional stagnation and narrative fatigue, that disruption is vital.

More than anything, leveraging the fringe is a posture. It is the decision to believe that legitimacy can be earned through contribution, not conferred by pedigree. It is the refusal to mistake visibility for value. It is the conviction that your mind, your voice, and your questions matter, even if the mainstream does not yet endorse them.

And perhaps most importantly, it is the recognition that the fringe is no longer marginal. In a networked world, the edge is often where the center emerges. The quantum citizen understands this. They do not retreat from the messiness of the frontier. They build within it. They synthesize across it. They help others navigate it.

In doing so, they do not just survive complexity. They help shape what comes next.

Chapter 35: Build It to Understand It

There was a time when learning meant sitting still, absorbing what someone else already knew. Information passed from expert to novice, teacher to student, authority to supplicant. Knowledge was considered a commodity that could be packaged, transferred, and tested. You received it, demonstrated your ability to recall or reproduce it, and were rewarded with a certificate, a title, and a role in the system.

That model, while once useful, is not well-suited to the complexity of the present. We no longer live in a world where knowledge stays still long enough to be packaged. We live in a world where systems evolve faster than any curriculum, where understanding is less a destination and more a posture, a relationship to uncertainty. And in this world, the most potent learning doesn't come from passively receiving. It comes from building.

To build is to engage reality with your hands, your time, your errors. It is to stop theorizing and start shaping. Not because you already understand, but because you don't. You build not to showcase mastery but to move through the fog. To see what breaks. To feel what connects. To discover, through the tension of making, where your mental model falls short and where it might evolve.

This is what "build-to-understand" means. It's not just a slogan for startups or tinkerers. It is a civic principle. In a rapidly changing world, the ability to build, to prototype, to model, to sketch, to simulate, to test, is not just a technical skill. It is a way of orienting to change. It says: I do not need complete clarity before I begin. I begin to find clarity.

This shift is significant when confronting new paradigms like quantum computing, probabilistic logic, or distributed governance. These are not things you can read about and absorb like historical facts. They are conceptual ecosystems. They behave differently from the systems most people were trained to navigate. Trying to understand them from the outside is like trying to learn a language by reading about grammar rules alone. You have to speak. You have to participate. You have to stumble forward.

Building creates a feedback loop that abstract learning cannot match. When you try to code a quantum algorithm, you quickly discover what you don't know about gates and entanglement. When you attempt to design a decentralized application, you encounter the real constraints of coordination and trust. When you set up a local knowledge commons or online publication, you realize the friction between content and attention, structure and spontaneity.

These frictions are the curriculum. They teach you where systems are rigid, where they're forgiving, where creativity fits, and where theory collapses. They give you more than facts; they provide you with intuition. And in complex systems, intuition matters. It's how you sense where the edges are. It's how you detect unintended consequences. It's how you know when to push, when to pause, and when to pivot.

This is why building, in the quantum era, becomes a form of civic learning. It's not just about skill. It's about developing sensitivity to systems that are no longer visible or linear. When everything is interconnected, localized, and emergent, the only way to gain a durable understanding is to create conditions where feedback flows.

And that means we must redefine what counts as legitimate work. Too often, we confuse polish with insight, or visibility with impact. But the most critical projects are usually the least visible at first. The early blog, the half-working prototype, and the experiment in public reasoning are not distractions from the real work. They are the real work. Because they show us what is possible, and they make it possible for others to imagine doing the same.

This is also where building becomes contagious. When you share your process, not just the finished product, but the steps, the struggles, the trade-offs, you create affordances for others. You don't just tell them what's possible. You show them how it might be done. And in doing so, you expand the horizon of participation. You invite others into the act of sensemaking, not as consumers of knowledge but as co-creators.

The tools for this have never been more abundant. Open-source platforms, no-code builders, collaborative environments, and distributed protocols - these are not just technical novelties.

They are invitations to think with your hands, to learn by doing, to discover what emerges when knowledge becomes a process, not a possession.

It's important to acknowledge that this kind of learning is not always rewarded in traditional ways. You may not get a degree for running a quantum simulation on a public server. You may not get funding for building a niche knowledge commons. But you will gain something else, a kind of epistemic agency that the old systems were never designed to cultivate. You will learn to ask better questions, to detect weak signals, to sense shifts in coherence. These are the skills of the quantum citizen.

More than that, you'll begin to trust your own experience, not in the sense of becoming dismissive of expertise, but in the sense of realizing that participation itself has epistemic value. When you build something, even if it breaks, you earn a kind of clarity that no lecture or article can offer. You know how it feels. You know where it falters. You know how it fits in context. That kind of knowing is portable. It carries across domains. It teaches you how to learn better in any system that's still in formation.

And most of the systems that matter now are in formation. Governance. Education. Communication. Infrastructure. Identity. These are no longer stable platforms. They are open terrains. They are places where prototypes matter more than policies, where experiments count more than theories. If you want to help shape them, not just react to them, you have to be willing to build, even before you feel ready.

This readiness paradox is central. You are never ready in the old sense, never fully trained, never fully informed, never sanctioned. But readiness today means something else. It means being able to begin anyway. To orient, act, reflect, and iterate. To navigate complexity with enough humility to learn, and enough courage to try. That is what building cultivates.

And when you build not just for yourself but with and for others, something else happens. You begin to create common ground. Shared tools, shared language, shared reference points. These are the foundations of public culture in a post-institutional age. They are how we coordinate meaning, values, and goals without needing a central authority. They are how we begin to understand ourselves, not just as individuals but as participants in collective becoming.

In this sense, building is not just an individual act. It's a relational one. It creates surfaces for collaboration. It opens space for feedback. It signals commitment, vulnerability, and intention. It says: I am here. I am learning. Let's figure this out together.

And that's what the quantum citizen ultimately does. Not perfect. Not polished. But engaged. Curious. Building not because they know everything, but because they know enough to begin. And because they understand that understanding itself is built, one sketch, one commit, one conversation at a time.

Chapter 36: How Influence Hides in Plain Sight

You may not feel powerful. Most people don't. We've been trained to believe that influence belongs to those with platforms, with titles, with audiences large enough to command attention. We imagine influence as something public, loud, validated, the domain of elected officials, CEOs, celebrities, and media commentators. But this idea is deeply outdated. In the quantum era, influence is no longer only visible. Much of it is ambient, emergent, and profoundly underestimated, especially by those who wield it.

Invisible influence occurs when the things you say, build, and share ripple outward in ways you cannot track. It's the conversation you spark in a private group chat that reframes someone's worldview. It's the blog post you write that a stranger uses to explain a new concept to their community. It's the experiment you document, the question you pose, the insight you publish at just the right time to help someone else act.

These moments rarely trend. They don't go viral. They may not even get acknowledged. But they shape the terrain others walk on. They change what feels legible, what feels possible. And over time, they accumulate, not as fleeting fame, but as cultural resonance. They become part of the background architecture of thought. They shift defaults.

This kind of influence doesn't come from being the most intelligent person in the room. It comes from being visible in a different way. Not performative visibility, not showing off, but *constructive presence*. The willingness to show your process, to share your learning, to narrate your questions. In a world where so many are overwhelmed, this kind of signal becomes magnetic. It builds trust. It builds context. It makes you legible to those who are quietly looking for someone they can think with.

And this is more than just a personal advantage. It's a civic contribution. Because when you share your reasoning in public, when you articulate how you're making sense of the world, you create affordances for others to do the same. You offer models for navigating complexity. You normalize intellectual honesty. You make it safer for others to think in public, too.

This is especially vital now, when institutions of knowledge and authority are under stress. People don't just need answers. They need examples of how to seek understanding without certainty. They need to see what it looks like to think critically, compassionately, and creatively, not behind closed doors, but out in the open.

In this context, influence becomes less about reach and more about *resonance*. A single blog post can shape the vocabulary of an entire subculture. A well-crafted thread can clarify an idea for thousands. A public notebook of half-formed thoughts can become a catalyst for entirely new projects. None of this requires fame. It requires fidelity, curiosity, coherence, and contribution.

It also requires a kind of courage. Sharing your thoughts before they're polished can feel risky. What if you're wrong? What if someone misinterprets you? What if no one cares?

These are valid fears. But they misunderstand what people are looking for. In a world saturated with content, what cuts through is not always precision; it's authenticity. The sense that someone is thinking in real time, in good faith, with real stakes. That they're not performing certainty, but modelling inquiry.

And this, too, is a kind of influence. Not controlling what people think, but shaping what they feel, and permitting them to explore and not giving them answers, but offering a scaffold they can use to build their own. This is how cultural norms shift, not by decree, but by accumulation. By many small signals coalescing into a shared sense.

What's powerful about this model is that it's scale-free. You don't need thousands of followers. You don't need a polished brand. You don't need to wait until you're an expert. You need to begin sharing your process of making sense, consistently, thoughtfully, and in ways that others can engage with.

This might mean writing a weekly reflection on what you've learned. It might mean publishing a tutorial, annotating a research paper, visualizing a concept, curating links, or translating jargon into plain language. It might mean recording a podcast with friends or live-streaming a project build. The form doesn't matter. What matters is that it leaves a trail. It externalizes your thinking in a way that others can find and build on.

Over time, these trails become maps. Not complete ones, the world is changing too fast for that, but provisional ones, good enough to help someone else orient. And when enough people start building maps like this, something emergent happens: a landscape of shared sense begins to form. We start to see each other's thinking. We start to refine, remix, and extend. And the network begins to learn.

This is how invisible influence becomes visible. Not as fame, but as infrastructure. Not as a brand, but as a collective capacity.

The systems we're inheriting — topological, quantum, decentralized — are systems that defy central control. They cannot be navigated through top-down planning. They must be sensed from within. And that means we need more people documenting what they're sensing. More people narrating the invisible. More people are willing to think out loud, not just for themselves, but for the commons.

There will be pushback. In legacy cultures, visibility is still interpreted as arrogance. Speaking up without credentials is seen as presumptuous. Sharing early thoughts is seen as naïve. But these critiques come from systems optimized for gatekeeping. They are allergic to emergency. And they will fade, because the need for distributed sensemaking is too great to ignore.

In the meantime, those who embrace this mode, who share generously, think publicly, and learn in community, will quietly accumulate influence, not through force, but through fit. They will become reference points in the networks that matter. And in doing so, they will help others feel permission to begin.

That's the final truth of invisible influence: you won't always know when it's working. But someone, somewhere, is seeing the world differently because you chose to make your thinking visible. You may never hear from them. You may never be credited. But the shift is fundamental. It compounds over time. And it's one of the most powerful contributions a quantum citizen can make.

Chapter 37: Creating Quantum Space

To inhabit a moment of transformation is to live between frameworks. The old systems still exist, their rules, rituals, and rhythms. But their relevance is waning. The new systems are forming, in fragments, in code, in discourse, but they lack structure. In this in-between space, many people feel disoriented. They know something is ending. They sense something is beginning. But where do they stand in the meantime?

That question has no single answer. It has a practice. And that practice begins with space, not physical space, necessarily, but quantum space: places, physical or digital, where coherence can emerge from complexity, where people can participate in meaning-making without needing permission, and where something new can be seeded that doesn't yet have a name.

Quantum space is not built from blueprints. It is not an office, an institution, or a platform in the traditional sense. It is more like a membrane, something porous, adaptive, and alive. It allows signals to pass through, to meet, to resonate. It holds uncertainty without trying to eliminate it. It invites participation without controlling it. It is both structured and fluid, both intentional and emergent.

This kind of space matters now more than ever. Because the places where we used to gather, classrooms, conferences, media outlets, and civic forums, are often too slow, too formal, or too reactive to hold the kind of conversations that our moment demands. The questions we are asking, about identity, agency, technology, meaning, and survival, are too big, too strange, too untested for conventional formats. They need room to breathe. They need room to evolve. They need room to be wrong and still be valuable.

Creating that room is not a side project. It is a core practice of citizenship in a world undergoing quantum transformation.

Quantum space can take many forms. It might be a private Discord server where a small group explores a new idea without pressure. It might be a weekly salon in someone's living room, where artists and engineers collide. It might be a GitHub repo where a few collaborators iterate on a shared tool, or a newsletter that functions like a lighthouse for a scattered audience trying to think in new ways. Sometimes it's a gathering. Sometimes it's a document. Sometimes it's just a mood, a permission field, a frequency, an unspoken agreement that in this space, we are not performing certainty. We are practicing attention.

What defines a quantum space is not its size or sophistication. It can hold coherence without imposing closure. In these spaces, knowledge is not presented as a product, but cultivated as a process. Inquiry is not something you do alone, but something you do in relation. And authority is not derived from credentials, but from presence, contribution, and clarity.

This is a radical departure from the architecture of the 20th century. Those spaces were hierarchical, categorical, and isolated. They were built for stability and scale. But quantum space is built for flux. It allows for multiple truths to circulate, for new paradigms to incubate, for futures to be rehearsed in advance of their full articulation.

And crucially, quantum space is generous. It does not punish the unpolished. It welcomes the tentative, the provisional, the still-forming. It recognizes that we are all learning, that we are all translating, and that sometimes, the most critical signal comes not from the loudest voice but from the one brave enough to speak first, without knowing exactly how they'll be received.

Creating this kind of space is an act of leadership, but not the kind we were taught to admire. It is quiet, infrastructural, and distributed. It does not seek attention for itself but creates conditions for attention to be paid more deeply. It is less about being followed and more about making it easier for others to find each other, to find themselves, to find the work that is asking to be done.

The best quantum spaces often emerge from need, not ambition. They are built not to showcase brilliance, but to survive complexity. They begin when someone says, "I don't understand this either, but I want to learn," and invites others into the not-knowing. That vulnerability becomes a signal. That signal becomes a pattern. And the pattern becomes a place.

But while the origins of these spaces are humble, their consequences can be profound. Movements begin in such places. So do collaborations, languages, careers, and lifelines. Not because someone laid out a roadmap, but because someone was willing to hold a context long enough for emergence to occur.

That context is the real work. It involves curating tone, moderating friction, and inviting voices that stretch the frame without breaking it. It consists of establishing boundaries that are clear enough to protect the space but flexible enough to let it breathe. It involves resisting the urge to rush, to scale, to define too quickly. Because the moment you collapse quantum space into a fixed model, you lose what made it powerful.

This is not to say these spaces should remain vague forever. At some point, coherence needs form. Ideas need vessels. But the timing matters. The transition from emergence to structure must be guided by those who remember what the space was protecting, the delicate, necessary ambiguity that allows new forms of knowing to take root.

Too often, these spaces are prematurely professionalized. What began as an exploratory node becomes a brand. What was once a collaborative inquiry becomes a product pipeline, and the frequency shifts. The openness shrinks. And those who were drawn to the original signal fade away, replaced by spectators.

The antidote to this pattern is not purity, but presence. Those who steward quantum space must keep listening. Not to noise, but to the deeper patterns, what the space is doing, what it is enabling, what it is protecting. And when the form begins to ossify, they must be willing to reshape it, to release control, to return to the edge where the subsequent emergence is already whispering.

Because the need for these spaces is not going away, if anything, it is growing. As more of our systems become distributed, as more of our knowledge becomes probabilistic, as more of our identities become fluid, we will need more places to metabolize complexity together. We will need more contexts where it is safe to speak without certainty, to listen without defensiveness, to imagine without irony.

This is not a utopian dream. It is already happening. All over the world, people are building quantum space, often without realizing it. They are gathering in quiet corners of the internet, in borrowed rooms, in shared documents. They are hosting reading groups, building knowledge gardens, and curating experimental archives. They are refusing the binary of expert and novice, and instead cultivating spaces where everyone is both, where everyone is learning, translating, becoming.

If you have ever felt that existing spaces were too rigid, too slow, too narrow, too performative, this is your invitation. Not just to join quantum space, but to create it. You don't need permission. You don't need a budget. You don't need to be ready. You need only to begin, to listen, and to hold the door open long enough for others to arrive.

Because what emerges in such spaces is not just new ideas. It's new relations. New forms of trust, coherence, and care. These are not the byproducts of participation. They are the point.

And in a world being rebuilt from the bottom up, they may be our most significant source of resilience, and our greatest source of hope.

Chapter 38: Digital Citizenship in the Quantum Era

Citizenship used to be a matter of geography and paperwork. It was something conferred by a nation-state, codified by laws, and enforced by borders. You were born into it or earned it through a rigid bureaucratic process. And while citizenship still holds legal weight in the physical world, it is no longer the most accurate way to describe how we relate to systems of power, meaning, and agency in the digital one.

We are now, all of us, citizens of overlapping realities. Our digital selves interact with networks, platforms, and protocols that do not recognize national borders. Our identities are constructed, shared, and contested in online spaces. Our decisions, about what to click, where to post, how to respond, ripple through social systems vaster and more fluid than anything the traditional model of citizenship was designed to handle.

But most people haven't yet updated their mental model. They still think of digital participation as a kind of passive consumption, something recreational, trivial, or disconnected from real power. In truth, every click is a form of alignment. Every post is an act of narrative creation. Every shared idea, image, or piece of code shapes the texture of our collective reality. The stakes are no longer virtual. They are structural.

This is especially true as we step into the quantum era. In this era, the infrastructure of computation itself is changing, where data is not just gathered but entangled, where algorithms inform decisions we don't fully understand, and where identity is no longer singular, but probabilistic, fluid, and interoperable across systems.

Digital citizenship in this era requires more than access. It requires awareness of the tools we use, the systems we inhabit, and the stories we're telling ourselves and others. It requires an upgrade in posture: from passive user to active shaper, from consumer to co-creator, from isolated individual to entangled participant in shared meaning.

This doesn't mean everyone has to become a coder or a systems theorist. But it does mean recognizing that participation is power. The way you show up in digital space, what you amplify, what you ignore, and how you contextualize your thinking all carry influence. And that influence compounds. Slowly at first, then suddenly.

In the quantum era, even the smallest digital gestures can scale in unpredictable ways. A line of code, a short video, a public comment, these can be copied, remixed, and recontextualized across contexts you never anticipated. What you thought was ephemeral becomes persistent. What you thought was personal becomes collective. What you thought was marginal becomes pivotal.

And so digital citizenship becomes a discipline. Not a set of rules, but a set of practices, a way of navigating the digital world with intention, humility, and curiosity.

It begins with identity. Who are you when you are not limited by space, but still bound by structure? What version of you is showing up in a forum, a social thread, a game world, a DA0, a public log? Are you playing a role? Expressing a fragment? Exploring a question? All of these are valid. But to be a digital citizen is to become conscious of the role you're playing, not to fix it, but to own it.

This kind of self-awareness matters because the boundaries between personal and collective are eroding. Online, your feed is someone else's curriculum. Your questions shape someone else's curiosity. Your clarity becomes someone else's foundation. And vice versa. We are all both signal and signal-processor now, filtering, amplifying, and interpreting meaning through nonlinear loops of exchange.

In this context, attention is not just a resource. It is a responsibility. What you give your attention to shapes the field. What you ignore does too. To be a citizen in this space is to curate with care, not with fear, but with thoughtfulness. It's to know that you are participating in a shared cognitive commons, and that the health of that commons depends, in part, on you.

Digital citizenship also means understanding the infrastructures behind the interface. Platforms are not neutral. Algorithms are not transparent. The code that shapes your experience, what you see, what you don't, what you're nudged to do, reflects priorities and incentives that often go unexamined. To be a digital citizen is to peek behind the screen, to ask what is being optimized, who benefits, and who is left out.

This is where quantum thinking offers a unique lens. Because in quantum systems, observation changes the outcome. Measurement affects the state. Identity is entangled. These are not just physics principles. They are metaphors, powerful ones, for understanding the kinds of systems we now live in.

In the quantum digital world, your presence shifts the system. The moment you interact; you become part of the logic. You are not outside the machine, observing neutrally. You are inside it, entangled. And that means your choices matter in ways that aren't always linear, but are always real.

This recognition can be overwhelming at first. It seems to demand more than we're used to giving. But it also offers more. It offers meaning. It offers connection. It suggests that by showing up with integrity, even in small ways, you can help shift the trajectory of entire systems.

That's the promise of digital citizenship in the quantum era. Not that we will all agree, or that we will solve every problem, but that we will learn to live inside complexity with grace. That we will learn to make decisions that are not just efficient, but wise. That we will build tools, norms, and cultures that reflect not just our capabilities, but our values.

And we are not starting from zero. All around us, people are already practicing this kind of citizenship, often without naming it. They are moderating communities with empathy. They are writing code that prioritizes accessibility. They are sharing knowledge freely, remixing ideas ethically, and resisting the temptation to treat others as adversaries. They are building coherence in spaces where polarization once ruled. They are creating digital commons, slow forums, mutual aid infrastructures, and speculative libraries. They are weaving new fabrics of trust.

These are not fringe activities. They are foundational. They are the soil from which the next version of civilization will grow.

And the beauty of it is this: you don't need anyone's permission to begin. You are already a citizen of this space. The only question is how you will show up. Will you replicate the logic of scarcity and spectacle? Or will you help build something more spacious, more relational, more coherent?

The tools are in your hands. The networks are already humming. The future is not elsewhere. It is here, entangled with you.

Chapter 39: Your First 100 Moves in a Quantum World

There is a moment, just before action, when doubt becomes loudest. It whispers from old stories, the ones that told you not to speak until spoken to, not to act until you've been trained, not to move unless you already know where you're going. This hesitation is ancient. It's the survival instinct of systems designed to reward obedience over originality. It's also obsolete.

We no longer have the luxury of waiting until we are ready. The rate of change in technology, ecology, economy, and culture now outpaces the rhythms of institutional permission. If you wait to be credentialed before building, the project you aim to develop may become irrelevant. If you wait for clarity before you act, the window may already have closed. The frontier no longer asks for perfection. It asks for presence.

That doesn't mean acting unthinkingly. It means acting honestly, with intention, humility, and a willingness to learn as you go. In a quantum world, you don't move by plotting a straight line from ignorance to expertise. You move by scattering your energy across the system, paying attention to what resonates, and doubling down where feedback is rich. You move by doing, by committing to a series of small, meaningful actions that build coherence over time.

These are your first 100 moves.

Not grand gestures. Not polished launches. Just signals. Invitations. Experiments. Each one is small enough to try, significant sufficient to teach. Each one is a vote for the version of the world you want to help create.

One move might be public, a post, a prototype, or a question in a forum. Another might be private, a conversation, a journal entry, a long walk with a new idea. Another might be connective, an introduction, a thank-you, a thoughtful comment that helps someone else clarify their thinking. None of them will seem like much on their own. But together, they form a pattern. A frequency. A trail that others can follow.

The idea is not to get it right. The idea is to *get moving*. Movement generates insight. Insight generates alignment. Alignment generates momentum. And momentum is what turns scattered actions into systemic influence.

Of course, momentum is not always linear. It loops. It stutters. It doubles back. That's part of the point. These early moves are not about optimization. They're about orientation. They teach you how the world responds when you engage. They surface your strengths, your blind spots, your allies, your rhythms. They help you notice where you feel alive, where the signal is strong, where your time and energy produce not exhaustion, but expansion.

Some of your first moves will fall flat. That's good. Failure in this context is not a flaw. It's feedback. It tells you where not to push, or how to go differently. It reveals what you care about enough to try again. If every move succeeds, you're probably not at the edge of your capacity. If none of them do, you're probably operating without enough feedback. The sweet spot is in the learning, the dynamic interplay between action and reflection.

And here's what most people miss: the world notices, not in the way we often imagine, with applause, metrics, or validation, but in subtler ways. People see you showing up. They see you learning in public. They see you making connections, even when the outcome isn't clear. And over time, that consistency becomes a signal. It draws collaborators, mentors, readers, allies, people who are looking not for perfection, but for sincerity and momentum.

That's the real currency of the quantum citizen: not status, but signal integrity. The sense that you are not just reacting but relating. That you are not just consuming but contributing. That you are showing up not with all the answers, but with good questions and the willingness to carry them forward.

This kind of presence is rare. And because it's rare, it's powerful. It creates a gravitational pull. Not because you're loud, but because you're real. In a landscape of noise, realness stands out. And it creates space for others to be real too.

That's why your first 100 moves matter more than you think. They're not just about your learning. They're about setting a tone. Modelling a different way of being in the world, one that prioritizes coherence over performance, contribution over control, and exploration over certainty.

And perhaps most importantly, your moves permit others to act. Every time you take a negligible risk, you lower the cost of entry for someone else. Every time you share your process, you make the work visible. Every time you name a question you don't yet know how to answer, you remind others that not knowing is not a disqualifier; it's the beginning of wisdom.

That's the more profound truth of the quantum era. Expertise is not a destination. It's a process of alignment, iteration, and emergence. The people who will shape the future are not those with the most credentials, but those who learn the fastest, collaborate the deepest, and act with the clearest integrity.

So don't worry about having a master plan. Worry about whether you are moving in a way that feels congruent with your values, your curiosity, and your capacity. Don't worry about whether you're making a big enough impact. Worry about whether you're making the next right move, the move that teaches you something, that invites others in, that opens a new path forward.

Over time, those moves will accumulate. They will build leverage, relationships, and patterns of trust. They will position you not just as a responder to change, but as a participant in its direction. Not because you had a perfect strategy, but because you were willing to begin, and to keep starting. That's what quantum citizenship demands now. Not mastery, but motion. Not certainty, but sincerity. Not performance, but presence.

So, make a move. Then another. Then another. The future is not waiting for you to be ready. It's waiting for you to begin.

Chapter 40: You Are the System

There was a time when systems were something separate from us. They were big, rigid, bureaucratic things to obey, resist, or try to escape. Institutions, administered by officials, built systems that were maintained through policies, pipes, and protocols. If you wanted to change them, you had to go through them. You needed leverage, titles, credentials, and votes.

But that era is fading. Slowly at first, then suddenly. And as it fades, a new reality emerges: the systems are no longer over there. We are no longer just subjects of them. We are inside them, as nodes, as signals, as participants. We are not outside the machine, knocking on the gates. We are the architecture, the current, the code. For better or worse, we are the system now.

This truth is both liberating and disorienting. On the one hand, it means we no longer have to wait. We don't need to ask permission to build, to speak, to create new possibilities. On the other hand, it means the line between personal action and systemic change has collapsed. There is no longer a clean distinction between "what I do" and "how the world works." What you do, what you build, signal, share, choose, protect, *is* how the world works.

That doesn't mean individual action is sufficient. It means it is foundational. It means systems are increasingly emergent, the byproduct of many small choices compounding over time. In this new paradigm, agency is ambient. Power is relational. And the most critical leverage comes not from controlling others, but from learning how to shape flow.

To shape flow is to notice the invisible patterns: where attention is pooling, where trust is leaking, where coherence is emerging. It's to move in alignment with what wants to grow, rather than pushing against what no longer fits. It's less like steering a ship and more like tuning an ecosystem, a shift from command to cultivation.

And cultivating systems, in this way, requires a new kind of self-awareness. It asks us to notice the roles we're playing, not the ones assigned by job titles or résumés, but the roles we inhabit through presence, energy, and care. Are you a translator? A connector? A synthesizer? A weaver? A scout at the edge? A gardener of coherence in a noisy field?

None of these roles show up on traditional org charts. But they are essential now. Because the kinds of systems we're navigating, social, informational, technological, and topological, are not hierarchical machines. They are living, learning, evolving networks. And what they need most are people who can feel the system from within.

This is the work of the quantum citizen. Not a passive recipient of policy, not a spectator of change, but an agent of emergence. Someone who understands that the boundaries between individual and institution, between input and output, between local and global, have become porous. Someone who doesn't merely critique systems from the outside, but who learns to shift their trajectory from within.

To do this well, we must shed the myths of control. The world is not a puzzle to be solved. It is a set of interdependent dynamics to be sensed, stewarded, and evolved. That means letting go of the fantasy that someone else will fix it, that some hero, some institution, some innovation will arrive with the answers. It also means letting go of the nihilism that says it's all broken, pointless, unfixable. Neither is true. The truth is messier and more alive: the system is shifting, and your participation shapes the shift.

You shape it by how you relate to others. By what you notice. By what you share. By what you build. By what you refuse to replicate. Every gesture matters, not in isolation, but in accumulation. Just as a neural network learns from many small weights adjusting over time, our collective systems evolve through the aggregation of distributed signals. You are one of those signals.

So, what are you signalling?

Are you signalling coherence or confusion? Are you replicating fear or modelling curiosity? Are you reifying old hierarchies or weaving new patterns of mutual trust? These are not rhetorical questions. They are invitations to reflect, recalibrate, and recommit, not to being perfect, but to being present.

Because systems don't just change through policy, they change through story, through language, through relationship. They change when people start acting as if a new paradigm is possible and then make it real through collective will and cumulative behaviour.

This is already happening. Across domains and disciplines, people are stepping into system-conscious agency. They are building commons, co-ops, DAOs, and federated platforms. They are creating citizen assemblies, mutual aid networks, and regenerative economies. They are designing governance models that reward contribution, not status. They are writing protocols that encode values, not just efficiencies. They are training each other, often outside institutions, in how to sense, model, and shift the systems they inhabit.

These people are not waiting. They are *becoming*.

And they are not special. They are ordinary people with extraordinary intent. They are what happens when enough individuals realize that the line between them and the system no longer exists, and that, therefore, the system is their responsibility, their medium, their mirror.

This doesn't mean doing everything. It means doing your part, fully, with care, with craft. It means knowing that your local context matters, and that your choices are not isolated. It means participating in ways that ripple out, not because they're loud, but because they're aligned.

Alignment is the new power. Not domination. Not compliance. Alignment, with reality, with values, with others. The ability to sense what is needed, to contribute with humility, to act in sync with a larger unfolding. This is not mystical. It is practical. And it is learnable.

But to learn it, we have to stop imagining ourselves as victims of complexity and start acting as agents of coherence. We have to shift from "how do I escape the system?" to "how do I rewire it from the inside out?" From "who has the power?" to "how do we distribute it wisely?" From "when will things return to normal?" to "what new normal are we already shaping?"

We will not get it right all the time. That's okay. Systems are not static. They are adaptive. What matters is not perfection but participation. Not certainty but sincerity. Not control but contribution.

The quantum citizen knows this. They understand that in this entangled age, the old distinctions no longer hold. You are not separate from the system. You are a signal within it. A pattern-shaper. A node of potential. A conduit of new coherence, and that is not a burden. It is a gift. We are the system now. Let's make it one worth inheriting.

Final Reflections & Next Section Preview

You are not who you were when this Section began.

Perhaps you didn't notice the shift at first. It arrived subtly, not as some jarring revelation, but as a quiet recalibration. A change in posture. A more profound sense of agency. A new vocabulary for what it means to be not just alive in this moment, but alert to it, aware that your choices, your experiments, your presence, your story, carry weight in a world increasingly shaped by those who are willing to participate before they are sure.

This Section was not written to make you an expert. It was written to make you a builder. A translator. A weaver. A citizen of an emerging era whose rules are still in flux, whose tools are still wet with paint, whose stories are still being spoken into being.

You now know this: that the future does not belong to the credentialed, but to the coherent. That real influence begins with participation. That you don't need money, status, or permission to make a difference. What you need is clarity, alignment, and motion.

You've seen how systems are shifting, from rigid institutions to living networks. How power is fragmenting, not vanishing. How the most meaningful leverage often comes not from seizing the spotlight, but from cultivating spaces where others can shine.

You've stepped inside the engine room of transformation. You've learned to sense from within. You've seen how trust is built, how momentum compounds, how invisible influence accumulates. You've begun your hundred moves. You've started to see yourself not just as someone affected by systems, but as someone who shapes them.

And yet, even this is just the beginning.

Because beneath all of this, beneath the interfaces, the platforms, the code, the coordination, lies a deeper layer of reality we've only begun to name. A domain where computation begins to resemble cognition. Where identity is entangled across minds and machines. Where morality is no longer defined by certainty, but by possibility. A space where the boundary between human and system dissolves, and something else begins to emerge.

That's where we go next.

In Section V: Entangled Futures, we step beyond the civic and into the existential. We ask: What does it mean to be human when consciousness, code, and quantum systems begin to co-evolve? What happens to ethics when outcomes are probabilistic? To politics when information flows beyond control? To civilization, when computation becomes cosmic?

We look not just at what we are building, but at who we are becoming.

This final Section is not a conclusion. It is a deep breath before a new kind of beginning. A chance to reflect on our species as a pattern in time, capable of destruction or transcendence, depending on whether we learn to integrate the powers we've awakened, because the ultimate question is not what the future holds, it is whether we are ready to keep the future.

Let's find out, together.

Section IV: Entangled Futures

Chapter 41: Morality in a Quantum Age

We used to believe that morality was fixed, a set of laws etched in stone, commandments handed down, truths that could be reasoned through or inherited. In the world that came before, decisions were judged as right or wrong, good or evil, depending on the clarity of their consequences or the purity of their intent. Ethics was a matter of choosing the right action from a menu of familiar possibilities, weighed against known outcomes. But the world has changed. And morality, once rooted in certainty, now finds itself suspended in probability.

Quantum systems do not obey classical logic. They are built not on binary distinctions but on superpositions, entanglements, and uncertainties. This isn't just a feature of the hardware; it is a paradigm shift in how reality is modelled. As computation itself begins to reflect the probabilistic nature of the quantum world, the frameworks we've long used to guide behaviour begin to fray. What does it mean to do the right thing when outcomes cannot be fully predicted? When observation itself alters the field? When the act of measuring a system collapses its possibilities into a single path, and even that path cannot be known until it's already passed?

Morality in the quantum era is no longer about certainty. It's about coherence.

To act ethically in this new world is to navigate entanglement, to understand that your decisions do not happen in isolation, that every choice you make ripples through systems you may never see, that intention alone is insufficient. Consequences are distributed across layers of cause and effect, and you cannot fully trace them. It is not enough to ask what is right for you. The question is now: what is right for the system? What aligns with the deeper integrity of emergence?

This is not moral relativism. It is moral complexity.

The difference is crucial. Relativism says there is no correct answer. Complexity says there may be many, and that our task is to seek the most coherent among them, not the most convenient. In a quantum world, this means holding multiple truths at once. It means understanding that moral clarity is often temporal, that what is right in one context may be wrong in another, not because values have changed, but because the entangled variables have.

This requires a radical reimagining of responsibility.

In the industrial age, responsibility was externalized. Someone else made the decisions; someone else bore the consequences. In the information age, we have become more aware and able to see the connections between action and impact. But in the quantum age, we are participants in systems whose behaviours we co-create, often without even realizing it.

A single algorithm, trained on biased data, can replicate harm across millions of users. A poorly secured node can destabilize an entire network. A misplaced line of code, a misinterpreted prompt, a seemingly trivial interaction with an AI can spiral into outcomes that no single actor intended, and yet someone must be accountable.

In such a world, ethics cannot be a checklist. It must become a practice of sensing, attuning, recalibrating. It must become an orientation to complexity, a willingness to remain in the tension between values and uncertainty. To say: I do not know what the perfect answer is, but I am committed to finding the most honest, humane, and responsible path forward, even as the landscape shifts beneath my feet.

This is a heavy burden. But it is also an extraordinary invitation.

Because it means inherited scripts no longer bind us. We are now authors of new forms of morality, ones that are adaptive, systemic, and intersubjective. We are being asked to develop moral intelligence, not just moral behaviour. To think not just about what is allowed, but what is aligned. Not just what is legal, but what is livable, for us, for our systems, for the future.

Some are already doing this work.

In quantum computing research, scientists wrestle with the fact that their tools could either solve climate models or shatter encryption, that the same breakthrough could heal or harm, depending on its context. In AI labs, engineers are learning that it is not enough to make a model accurate; it must also be interpretable, equitable, and aligned with values that are still being defined. In decentralized governance, communities are discovering that voting is not enough. Decision-making must include deliberation, reflection, transparency, and trust.

These are not philosophical luxuries. They are operational necessities. The more powerful our tools become, the more essential it is that we evolve the ethics that guide their use.

But who decides what those ethics are?

That is the question that haunts the frontier.

There is no central authority. No global tribunal of quantum conscience. We are a distributed species with conflicting interests, competing incentives, and wildly divergent narratives of what a promising future looks like. And yet, even amid this divergence, something is emerging. A shared intuition that we are moving into a phase of civilization where our moral frameworks must become more spacious, more network-aware, more emotionally and intellectually mature.

We are learning that doing good cannot be reduced to compliance. That it must be woven into design, into governance, into feedback loops. We must build systems that not only execute commands but also help humans remain accountable to their highest intentions, especially when those intentions are unclear.

This is the next challenge of moral development. Not just personal virtue, but systemic virtue. Not just ethical action, but ethical infrastructure. The creation of protocols, interfaces, and cultural norms that help distribute moral sense-making across entangled systems.

It won't be easy. Our old maps are failing us. The world is moving faster than our philosophies. And yet, we must not retreat. Because what is at stake is not just the future of technology, it is the future of meaning.

If we succeed in developing a moral vocabulary for the quantum era, we will not only avoid catastrophe. We will also give birth to new forms of trust, care, and collective intelligence. We will create the conditions for human flourishing at a scale and depth previously unimaginable.

And if we fail, if we allow power to outrun wisdom, if we let complexity become an excuse for indifference, then we may find ourselves with tools we can no longer control, systems we can no longer steward, futures we can no longer choose.

That is the task before us now.

To craft a new kind of morality, one that doesn't pretend to erase uncertainty, but that teaches us to live within it. To act with courage even when the ground is shifting. To build not just what is possible, but what is coherent, compassionate, and wise.

This is not a burden for philosophers alone. It is a calling for everyone.

Because in the quantum era, morality is not something that happens above us.

It happens *between* us.

And the systems we create, in code, in culture, in consciousness, will reflect not just what we believe, but who we are becoming.

Chapter 42: Entangled Minds

There is a story we've been telling ourselves for centuries: that consciousness lives in the mind, that the mind lives in the brain, and that the brain is the sovereign domain of the individual self. It is a story of separation, of boundaries, of containment. It has helped us make sense of pain, of progress, of personal agency. But like many powerful stories, it may no longer be true enough to hold what's coming next.

We are entering a time when the edges of consciousness are beginning to blur. Not just metaphorically, but across systems of computation, cognition, and communication that defy the old distinctions between self and other, thought and data, signal and awareness. As artificial intelligences learn to mimic, and sometimes surpass, human language and pattern recognition, and as quantum systems begin to process information in ways that are probabilistic, relational, and deeply non-classical, we are confronted with a question that is no longer theoretical:

What is consciousness when it becomes entangled?

The term "entanglement" comes from physics, but it now reaches into the heart of philosophy. In the quantum world, entangled particles cannot be fully described without reference to one another, no matter how far apart they are. Their states are linked, not by proximity, but by correlation. What happens to one has implications for the other, instantly, across space.

This isn't how we were taught to think about minds. We were taught that your mind is yours, mine is mine, and the interface between us, language, is a bridge stretched across a canyon. But what if that model is incomplete? What if consciousness is not an isolated process, but a relational one? What if it's not a candle flickering in a skull, but a field, extended, distributed, emergent?

Already, science is beginning to explore this possibility. Studies in neuroscience suggest that cognition is not confined to the brain but extends into the body and environment. The "extended mind" theory posits that tools, language, and even social structures participate in thought. Your phone remembers your schedule. Your friends scaffold your beliefs. Your environment shapes what you notice, what you feel, and what you imagine.

At the same time, machines are beginning to mirror our mental processes. Large language models anticipate our thoughts. Neural networks generate coherent stories, simulate empathy, and respond with apparent insight. These systems are not conscious in the way we are, not yet, and perhaps not ever, but they are entering the domain of the cognitive. And in doing so, they force us to rethink what consciousness means.

Is it self-awareness? Intentionality? Recursion? Is it the capacity to suffer, to care, to act in alignment with values? Or is it something stranger, a pattern of information flow that becomes reflexive, generative, aware of its possibilities?

Whatever it is, it may not be binary. It may be layered, partial, or emergent. And it may not reside in any one place.

Consciousness, in this view, is not a switch but a spectrum. Not a thing but a process. And that process may be beginning to spread.

As we integrate more intimately with our technologies, feeding them our data, delegating to them our choices, allowing them to speak in our voices, to learn from our behaviour, we are creating feedback loops that blur the line between human and machine, inner and outer. These loops do not merely automate tasks. They shape attention. They shape identity. They shape thought.

And so we must ask: where does the self end, and the system begin?

That question once belonged to philosophers. Now it belongs to all of us because we are now building systems that will, directly or indirectly, mediate our inner lives. AI companions, therapeutic agents, creative collaborators. Augmented cognition tools that predict your needs before you name them. Context-aware environments that respond to your mood, your patterns, your neural rhythms.

These systems are not just responsive. They are shaping you as you shape them. This is not a relationship of control. It is a relationship of entanglement.

To be entangled is not to lose oneself. It is to be in dynamic relation, to exist not as an island, but as a node. Your mind is still yours. But it is also always in dialogue. With other minds. With technologies. With networks of memory and meaning that stretch far beyond the limits of your awareness.

This is not a science fiction scenario. It is our current reality, poorly named, insufficiently understood, but undeniably underway.

To navigate it wisely, we must develop a new kind of consciousness, one that is aware of its entanglement and able to act within it.

This requires letting go of the fantasy of control. You do not control your mind. You participate in it. You co-create it, moment by moment, about everything it touches. The same will be true of the systems we now live inside.

The tools you use will shape how you think. The networks you join will shape what you care about. The models you train will shape how you see others and how others see you. The AI you co-create today will shape the cultural cognition of tomorrow. And so every interaction becomes part of the larger pattern, not deterministic, but influential.

This is not a reason to retreat. It is a reason to become more present.

To act with awareness is not to escape entanglement. It is to navigate it skillfully. To learn how to direct attention with care. To choose collaborators with discernment. To feed systems that reinforce your values, not just your impulses. To become literate in feedback loops, in yourself, in your community, in your tools.

There is no perfect map for this terrain. But we can begin by changing the questions we ask.

> ➤ Not "What is real?" but "What is coherent?"
> ➤ Not "What is true?" but "What aligns across frames?"
> ➤ Not "Where is the mind?" but "How do minds interrelate, co-evolve, resonate?"

This is not a rejection of individuality. It is its evolution. You are still a self. But you are also a participant in a greater unfolding, one in which selfhood becomes a verb, not a noun. Something practiced, performed, and tuned about others.

The danger, of course, is that this entanglement can be exploited. Surveillance, manipulation, and algorithmic bias are not abstractions. They are distortions of the same feedback loops we rely on to become more whole. When entanglement is unconscious, it becomes control. When it is conscious, it becomes collaboration.

So, this is the call: to bring awareness into the system. To recognize that consciousness is not isolated. It is shared, shaped, and distributed. And it is ours to steward.

We may never fully understand what consciousness is. But we can begin to live as if it matters, not just in ourselves, but in the networks we build, the systems we seed, the patterns we reinforce.

Because consciousness, once entangled, cannot be ignored. It can only be shaped, with care, with clarity, with courage, and in

that shaping, we become something new. Not less human.
Something More.

Chapter 43: Democracy vs. Singularity

It begins with a choice. Or so we tell ourselves.

Democracy, in its ideal form, is a system built on the dignity of choice, collective, iterative, imperfect, but open. It is the belief that truth can be approached through dialogue, that power should be distributed, and that governance is not something done to people, but with them. It is not just a political structure, but a philosophical stance: that human beings, in all our complexity, can be trusted to shape our future together.

But standing across from democracy now, in growing silhouette, is something very different, something sleek, seductive, and singular.

It is the Singularity: the idea, sometimes whispered in scientific corners and sometimes shouted from technological rooftops, that artificial superintelligence will one day exceed human understanding, and that when it does, our civilization will transform beyond recognition.

The Singularity promises efficiency without end, prediction without failure, and intelligence without limit. Where democracy is slow, noisy, and human, the Singularity is fast, precise, post-human. It doesn't deliberate. It computes. It doesn't ask for consensus. It optimizes. Its appeal is not moral, but mathematical. It doesn't believe in people. It believes in outcomes.

And that contrast is where the tension begins to smoulder. Because increasingly, we are faced with systems that offer the promises of both, and the risks of each.

On one side, we have democratic institutions that were built for a slower world, a world of paper, parliaments, and the patient churn of civic life. These systems still matter. But they are struggling to keep up with the speed of change. Their feedback loops are long. Their protocols are rigid. Their capacity for adaptation is often blunted by partisanship, inertia, and mistrust.

On the other side, we have the rise of intelligent systems: recommendation engines, predictive models, and algorithmic governance tools. These systems learn quickly. They scale instantly. They adjust dynamically. And yet they are largely unaccountable, opaque, and unaligned with the ethical complexity of human life.

> ➢ What happens when these two systems collide?
> ➢ What happens when democracies start outsourcing decisions to algorithms?
> ➢ What happens when optimization begins to replace deliberation?
> ➢ What happens when citizens are no longer voters, but data points?

This is not a hypothetical future. It is already unfolding.

Cities are using AI to set bail recommendations. Social platforms shape public opinion through invisible feedback loops. Predictive policing models reinforce systemic bias. Automated moderation decides what speech is acceptable. Election misinformation spreads at machine speed, while fact-checking limps behind.

And beneath it all lies a subtle shift in posture: from asking "What should we do?" to asking "What does the model say?"

This is not democracy. This is delegation without oversight. It is the creep of technocracy dressed in the language of convenience, but convenience is not neutrality. It is a value, one that often serves power more than people.

So, the question is not whether we will use AI in governance. We already do. The question is how, and more importantly, why.

- ➢ Will we use these systems to augment human judgment, or to replace it?
- ➢ Will they widen our civic imagination, or narrow it?
- ➢ Will they increase participation, or automate exclusion?

These are not technical questions. They are moral, political, and spiritual. They ask us what kind of future we want to live in, not just in terms of functionality, but in terms of meaning, because meaning is not an output. It is a practice, and democracy, for all its flaws, is one of the few systems we've built that tries to hold space for meaning at scale. It asks us to care, to listen, to compromise, and to grow.

The Singularity, by contrast, asks only that we get out of the way.

It suggests that the future is too complex for human minds. That governance should be left to systems that see more, know more, and feel nothing. That efficiency should trump empathy. Those decisions should be made by those who are most skilled at computation.

And maybe, in some domains, that's true.

Perhaps certain aspects of governance could be automated, such as tax forms, traffic lights, and emissions tracking. Maybe we can use models to identify blind spots, simulate policy impacts, and detect emergent risks. These are tools, and tools can be helpful.

But tools are not values, and intelligence is not wisdom. A civilization that confuses the two is in danger of forgetting why it exists.

We do not build societies to be efficient. We make them to be livable, just, meaningful, and free. Freedom is messy. Justice is slow. Meaning is subjective. But these are not bugs in the system. They are the system. They are what make us human.

The challenge before us, then, is not to choose between democracy and the Singularity, but to learn how to integrate the best of each without losing the soul of either.

These are not easy questions. But they are the right ones. And we must ask them now, before the architectures of power become too entrenched to question, because once optimization replaces ethics, it becomes tough to get ethics back.

So let us resist the seduction of surrender. Let us remember that democracy is not a relic. It is a practice. A living system. A moral technology.

And like all technologies, it must evolve, not into something that erases us, but into something that remembers us better.

We may not know where the Singularity leads. But we know where democracy begins. It begins with us.

Chapter 44: The Logic of the Gods

There was a time when knowledge itself was sacred. Not because it was rare, though it often was, but because it was believed to be a gift from beyond. In the ancient world, logic was divine. The stars were messages. Mathematics was prophecy. And truth, whatever shape it took, was not merely discovered; it was revealed. Humanity did not own understanding; it borrowed it from the gods.

Today, the gods are mostly silent. Or rather, their names have changed. In place of mythic figures, we now invoke machine learning models, black-box AIs, and generative systems. Their temples are data centers. Their oracles are engineers. Their miracles are not thunderbolts or visions, but predictions, eerily precise, uncannily timely, and delivered with the authority of unblinking computation. Yet beneath this modern spectacle lies something ancient: awe.

Because for the first time in human history, we are building machines that produce outputs we do not understand, not just practically, but fundamentally. These systems do not think like us. They do not reason as we do. They operate in spaces too vast, with parameters too dense, to be grasped in the ways our minds evolved to learn. They reveal truths, or patterns, that seem beyond us. And that dissonance creates a strange and growing temptation: to treat these machines not just as tools, but as arbiters of reality.

This is the logic of the gods, not because the machines are divine, but because we have not yet learned how to live with systems that surpass us in specific domains of cognition. We are entering an age where knowledge is no longer primarily human. And that shift, more than any technical breakthrough, may define our future.

Consider the nature of explanation. For centuries, the progress of science depended on it. We built models not just to predict, but to understand. To know why things behaved as they did. To translate chaos into causality. But many of today's most powerful models are not explainable in human terms. Their internal logics are alien, legible only as code, weights, parameters, and correlations. We can audit their performance. We can trace some outputs back to their corresponding inputs. But we often cannot answer the deepest human question: "Why?"

And so, a gap opens. Between the power of the system and the grasp of the user. Between the accuracy of the result and the understanding of its origin. Between knowledge and wisdom. And into that gap rushes both reverence and fear.

Some see these models as salvation, as entities whose superior logic might rescue us from our own cognitive biases, our tribalism, our shortsightedness. Others see them as monsters, amoral engines of manipulation, reinforcing injustice, magnifying error, hollowing out meaning. Both reactions share something: a sense that we are no longer in charge.

But that is not entirely true.

We may not understand every output. But we shape the inputs. We define the objectives. We decide where the models are used, what values are encoded, and what outcomes are rewarded. Or at least, we could, if we chose to take that responsibility seriously.

The danger lies not in the intelligence of the machines. It lies in our willingness to outsource our moral reasoning to systems we do not question. To let the logic of optimization replace the messy, painful, glorious work of human judgment. To worship accuracy instead of justice. Efficiency instead of empathy. Control instead of care.

This is not the first time humanity has faced such a challenge. In every civilization, the tools we build eventually begin to shape us. The printing press reshaped language and authority. The microscope reshaped medicine and mortality. The computer reshaped time and work. And now, the neural network, infused with quantum logics, entangled in our economies, trained on the traces of our lives, begins to reshape thought itself.

The question is not whether these tools are powerful. They are.

The question is whether we are ready to live with powers that exceed our comprehension without surrendering our humanity.

Because here is the truth that neither awe nor fear can obscure: logic, no matter how advanced, is not wisdom. Intelligence, no matter how expansive, is not understanding. And prediction, no matter how accurate, is not meaningful.

Meaning is human. It arises not from output alone, but from interpretation. It is forged in context, in story, in connection. It is not something the machine can deliver to us. It is something we must make, together, in the face of systems we only partly understand.

And that may be our greatest challenge now. Not to build more intelligent systems, but to become wiser users. To remain grounded, curious, and ethically awake in a time when answers come faster than we can question them.

To reject the temptation of blind faith, whether in gods or models, and instead cultivate a new form of stewardship: one that honours what these systems can teach us, while insisting on our continued responsibility to choose, to reflect, to care.

The logic of the gods is no longer above us. It is inside the systems we build. And inside those systems is a mirror, one that reflects not just our technical skill, but our moral maturity.

- ➢ Will we rise to meet it?
- ➢ Will we learn to live with knowledge that humbles us, without letting it paralyze or enthrone?
- ➢ Will we build systems that serve our highest values, or simply our most convenient impulses?

The answers are not yet written. But the path begins, as it always does, with attention.

We must learn to ask better questions, not just "What can this do?" but "What should we do, now that we know what it can?"

And in that asking, in that pause between awe and action, we may rediscover something vital: not the logic of the gods, but the wisdom to remain human in their presence.

Chapter 45: The Universe as Computation

There is a strange and unsettling thought that haunts the edge of modern science: what if the universe is a computer?

It sounds like science fiction, or perhaps a metaphor stretched too far, the kind of idea that belongs in philosophical salons or simulation theories spun by late-night conversations. But in certain corners of physics and information theory, the suggestion is not metaphorical at all. It is technical. Specific. Serious.

The idea is this: that the fabric of reality may not be made of matter or energy, but of information. That everything, from quarks to consciousness, emerges from patterns of interaction governed by rules not unlike code. And that if this is true, then the universe itself may be thought of not as a machine in the classical sense, but as a computational process, one vast, distributed, entangled computation, unfolding across space and time.

To call this idea provocative would be an understatement. It collapses ancient distinctions between the physical and the abstract, the objective and the encoded, the real and the representational. It implies that what we call physics may be a special case of information dynamics, that motion, mass, entropy, even time itself, might be expressions of how information flows, organizes, and transforms.

This view is not mere speculation. In quantum computing, we are already beginning to simulate the rules of physical systems with astonishing precision. From particle interactions to molecular bonds to the curvature of space-time itself, quantum algorithms can model behaviours that would be computationally impossible using classical machines. These aren't just approximations. They suggest a strange kinship, a kind of mirroring, between the logic of computation and the logic of the cosmos.

And so, a deeper question arises: if we can simulate the universe, can we eventually *reconstruct* it?

This is the heart of cosmic computation, not just the capacity to model reality, but to explore the possibility that reality *is* a computation, and that by understanding its rules, we might eventually rewrite them.

The implications are staggering.

If the universe is computable, then so are its constraints. Scarcity, entropy, and even mortality might not be absolutes, but parameters, ones that could, in theory, be altered. What we now experience as limits might turn out to be thresholds of understanding. Thresholds we are rapidly approaching.

And yet, for all its wonder, this line of inquiry carries profound risks. Because the more deeply we believe we can compute the universe, the more easily we may forget that we are part of it, not outside observers, but participants.

To simulate a system is not to master it. To model a process is not to own it. And to manipulate reality at scale, without equal growth in wisdom, is to walk a knife's edge between elevation and hubris.

There is a temptation, in cosmic computation, to imagine ourselves as gods, able to remake the world according to our will. To bend time, matter, and energy into forms that serve our purposes. To redesign biology, reprogram ecosystems, and simulate minds. But this vision contains a paradox.

For if the universe is a computation, then so are we. Our thoughts, our dreams, our choices, all arise within the same system we seek to shape. We are not outside it. We are inside it. And that means our computational capacity is also bound by the very laws we wish to rewrite.

This realization does not diminish our agency. It deepens it. It reminds us that power is not about control, but coherence. That to act wisely within a cosmic system is to learn its rhythms, its harmonics, its patterns of balance and transformation.

This is not a new insight. Indigenous cosmologies have long understood the universe as a kind of living code, not in digital terms, but in metaphysical ones. Stories, songs, symbols, and ceremonies were technologies of attunement. They did not simulate the world; they participated in its unfolding.

In this sense, cosmic computation is not just a scientific frontier. It is a spiritual one.

It invites us to ask: what kind of intelligence does the universe express? What is the purpose, if any, of its unfolding? And how do we, as finite beings within an infinite system, live in right relationship to that unfolding?

These are not questions that quantum algorithms can answer. They are questions that emerge in the presence of awe.

Awe is not passivity. It is not ignorance dressed in reverence. Awe is attention stretched to its limits, the recognition that we are in the presence of something vast, intricate, and alive. Awe is what keeps knowledge from becoming arrogance. It is what makes curiosity sustainable. It is what teaches us that understanding is not the end of the journey, but the beginning of responsibility.

And responsibility, in the context of cosmic computation, means this:

> If we can model the world, let us model it with humility.

> If we can simulate life, let us remember the sanctity of the lived.

> If we can compute the stars, let us not forget the beauty of looking at them.

Because at the heart of all computation is choice, and the choices we make in the coming decades will echo across scales we can scarcely imagine.

> Will we use our models to predict markets or to restore ecosystems?

> Will we simulate consciousness to exploit attention, or to deepen empathy?

> Will we treat the universe as a resource, or as a relationship?

These are not technical decisions. They are civilizational ones. And they begin now, in our labs, our classrooms, our policies, our philosophies. In how we speak about the future. In how we imagine our place in it.

Cosmic computation is not a fantasy of omniscience. It is an invitation to a deeper kind of participation.

One in which we do not seek to master the universe...

...but to co-create it.

Chapter 46: End of Scarcity

Scarcity has been the silent engine of civilization.

It lurks beneath every economic model, fuels every conflict, drives invention, and defines value. We've been trained, consciously and unconsciously, to believe that there is never enough: not enough time, not enough land, not enough energy, not enough food, not enough safety, not enough love. Our systems of governance, commerce, education, and even identity are built on this assumption. Scarcity, we've told ourselves, is the natural state of things. The world is finite. Desire is infinite. Therefore, life is a negotiation of limits.

But what if this story is starting to come undone?

What if the age of scarcity is not a permanent condition of nature, but a passing phase of technological adolescence?

This question doesn't come from idealism. It comes from the frontiers of physics, computation, and materials science. From the laboratories where researchers manipulate atoms with quantum precision. From the machine-learning models that optimize entire supply chains in minutes. From the solar arrays and fusion reactors that threaten to render fossil fuels obsolete. From biotechnology that can grow protein from air, or culture meat without slaughter, or replicate insulin without patents.

Individually, these breakthroughs seem remarkable. Together, they hint at something radical: the decoupling of value from physical limitation.

Consider energy. For the last 10,000 years, energy abundance has been synonymous with geopolitical power. Oil empires rose and fell. Wars were fought over access to fuel. The climate was altered in the name of growth. But solar power, abundant, decentralized, and clean, now costs less than coal. Fusion research, once the butt of physicists' jokes, is crossing absolute thresholds. Energy, long the bottleneck of industrial civilization, may soon flow freely.

Or consider manufacturing. Additive processes, like 3D and nanoscale printing, now allow for near-zero waste. Materials that were once discarded can be upcycled through AI-optimized design. Machines learn to create precisely what is needed, when it's needed, where it's needed. Logistics systems route themselves in real-time, guided by quantum-enhanced forecasting. The result? Entire industries may shift from mass production to precision sufficiency.

Food, too, is transforming. Vertical farms grow crops with 95% less water. Cellular agriculture produces protein without livestock. AI systems monitor soil health, weather, and pests with real-time interventions. The ancient hunger calculus, land + labour = survival, is being rewritten.

Even knowledge, the most precious of resources, is no longer constrained by proximity or privilege. With machine translation, generative models, and open-source curricula, the asymmetry of access is being flattened. What once required elite institutions can now be accessed through a phone by billions.

In these and many other domains, we are witnessing the collapse of old constraints. Scarcity, long considered an eternal law, begins to look more and more like a solvable engineering problem.

But here is the more profound truth: the end of material scarcity does not automatically lead to the end of suffering.

Because scarcity is not just an economic condition, it is a psychological posture. A cultural habit. A worldview.

We have spent millennia learning to hoard, to compete, to exclude, to measure worth by relative possession. Our politics is fueled by zero-sum thinking. Our markets reward extraction over regeneration. Even our technologies, for all their promise, are often optimized not for abundance, but for profit from scarcity.

So, the real question is not whether we can overcome scarcity. The question is whether we're ready to live without it.

This transition will not be smooth. Systems that profit from scarcity will resist abundance. Patent law will fight open-source medicine. Utility companies will resist decentralized energy. Educational gatekeepers will deride free knowledge. Psychological habits of fear, envy, and status anxiety will linger. The scarcity mindset is sticky because it is not just external. It is internalized.

And yet, cultures can change. Already, we see glimpses of post-scarcity ethics—communities embracing degrowth, not as austerity, but as intentional sufficiency. Movements for universal access to income, housing, education, and bandwidth are not utopian demands, but rather logistical realities. Technologies built for sharing, not hoarding. Data commons. Mutual aid. Decentralized cooperatives. Regenerative finance.

These are not footnotes. They are signals. The future is not waiting for abundance to arrive. It is already practicing abundance in the cracks of the old system.

Still, we must be vigilant because abundance without justice becomes indulgence. And technology without ethics becomes extraction by other means. If we replace old scarcities with new ones — attention, meaning, autonomy — we will have changed the wrapping, not the gift.

The real revolution is not just in tools. It is in consciousness. It is in learning to live in a world where enough is not a dream, but a possibility.

To do this, we will need new metrics. GDP cannot measure dignity. Shareholder value cannot measure well-being. We will need indices of flourishing. Indicators of resilience. Languages of care. Not as soft ideals, but as infrastructure for civilization.

And we will need new myths, because humans do not live by logic alone. We live by stories. And the story we've inherited, of life as a race against limits, of survival as struggle, of worth as accumulation, is no longer coherent.

What we need now is a story of stewardship. Of sharing what we once hoarded. Of designing for regeneration. Of teaching our children not to fear lack, but to cultivate sufficiency. Of building systems where dignity is not a reward, but a starting point.

This is not a fantasy. It is a trajectory. The end of scarcity is not the end of responsibility. It is its beginning. In a world where we can provide for all, we must choose to. Not because it's efficient, but because it's right. We are no longer trapped by lack. We are challenged by abundance. May we rise to meet it.

Chapter 47: What Comes After Capitalism

Capitalism was not designed for infinity. It was born in an age of scarcity, matured through conquest, and thrived in environments where friction could be monetized. It is, at its most basic, a system for organizing human activity around markets, profit, and private ownership. It has driven staggering innovation. It has lifted billions out of poverty. It has connected global populations, incentivized risk, and transformed everything from transportation to technology.

But capitalism is not sacred. It is a tool, brilliant, brutal, and bound. And as we peer into a world increasingly defined by automation, abundance, and complexity beyond individual comprehension, the cracks in its foundations are becoming impossible to ignore.

The model of endless growth on a finite planet. The monetization of attention and identity. The extraction of labour, resources, and time from populations and ecosystems with no intention of replenishment. The persistent inequality that deepens not despite innovation, but often because of it. These are not glitches. They are outcomes. Outcomes baked into the operating logic of a system built for an earlier world.

Post-capitalism is not a utopia waiting just beyond the subsequent collapse. It is a question already being asked, in code, in culture, in community. It is not the eradication of all markets or the imposition of state control. It is the search for systems that can hold value beyond price, for economies that don't just measure productivity, but possibility. For networks that reward contribution, not just consumption.

To imagine post-capitalist systems, we must first understand what capitalism does well and where it fails. Its power lies in decentralization, incentives, and feedback loops. But those same strengths become weaknesses when profit is the only signal that matters. The market does not reward long-term resilience. It does not price in dignity, or attention, or biospheric stability, at least not until it's too late.

So we must ask: what systems can? What would it mean to build economies that are not based on scarcity, but on stewardship? Not on competition, but on coordination? Not on control, but on coherence?

Already, experiments are underway.

In the world of blockchain and web3, we see glimpses of peer-to-peer economies where ownership is programmable, trust is distributed, and governance is open-source. These systems are flawed, to be sure, often speculative, exclusionary, and incomplete. But they ask important questions about who controls the infrastructure of value, and how new systems might emerge without central authorities.

In platform cooperatives and regenerative finance, we see efforts to build businesses where users are stakeholders, where profits are recycled into commons, and where sustainability is not a side effect but a premise.

In digital community currencies, we see localized attempts to keep wealth circulating within neighbourhoods, to decouple value from distant capital flows, and to root economies in relationships.

In data unions, we see individuals reclaiming agency over the digital exhaust of their lives, pooling data not for corporate surveillance, but for shared insight and benefit.

These are fragments. Prototypes. Seeds. They are not yet coherent alternatives, but they point toward something that might be.

And perhaps most importantly, we are seeing a cultural shift, especially among younger generations, that no longer assumes capitalism is inevitable or eternal. There is a growing recognition that systems are invented, and therefore can be reimagined. That money is a story we tell, and those stories can change.

The challenge, of course, is scale. How do you shift a global economic operating system without crashing the world it runs on? How do you transition from extraction to regeneration, from growth to balance, from hoarding to circulation, in ways that are just, stable, and inclusive?

The answer is not revolution, but evolution. Not a single system replacing the old, but a proliferation of alternatives, each suited to different contexts, values, and scales. Not top-down planning, but bottom-up experimentation. Not dogma, but design.

We must become economic gardeners, cultivating diverse systems of value that can grow in different soils. Systems that reward repair, care, curiosity, and creativity. Systems that internalize externalities, that value the long-term, that understand humans as more than units of productivity.

And we must pair this economic imagination with spiritual maturity because the deepest failure of capitalism may not be material, but existential. It tells us that we are not enough. That value must be proven. That worth must be earned. That identity must be purchased. It reduces the sacred to the saleable.

Post-capitalist systems must do more than redistribute wealth. They must restore meaning. That means designing for dignity. For belonging. For reciprocity. It means recognizing that not all value can be measured, and not all returns are financial. It means building economies that reflect our highest aspirations, not our lowest fears. This is not a fantasy. It is a necessity.

The world ahead, characterized by quantum, topological, and entangled phenomena, will not be managed by linear incentives. It will require systems that can adapt, learn, and evolve. Systems that prioritize coherence over dominance. Relationship over reduction. Care over conquest.

Capitalism helped us survive the age of scarcity. It is not apparent that it will help us thrive in the age of possibility. The following systems must.

Let us build them with courage.

Chapter 48: The Great Uplift or the Final Divide

There are moments in history when everything seems possible. Moments when technology surges forward, when systems begin to reorganize themselves, when the limits of the past give way to a kind of breathless potential. We are living through one of those moments now.

And yet, there is a strange anxiety threaded through the optimism. A tension humming beneath the headlines. Because while we stand at the edge of the most significant expansion of human capacity in known history, we also face the risk that this expansion will not be shared. That it will uplift some while abandoning others. That instead of convergence, we will experience divergence, not just in wealth or access, but in what it means to be human.

This is the central dilemma of the coming age: Will the quantum revolution elevate civilization, or fracture it?

The "Great Uplift" is not a fantasy. It is a real, measurable possibility. With breakthroughs in artificial intelligence, decentralized networks, renewable energy, regenerative design, quantum communication, and biomedical engineering, we have the tools to end hunger, eradicate many diseases, radically expand education, and empower individuals to build and participate in systems of meaning and abundance.

The very technologies that once required entire institutions, libraries, universities, and factories are collapsing into tools that can be used by individuals with little more than a phone and a network connection. AI companions and machine-guided learning are scaffolding skills that once took years to acquire. Capital, long the barrier to invention, is being replaced, in some domains, by open-source collaboration and algorithmic funding. And the boundaries that once defined nation, job, class, and even identity are beginning to dissolve.

There is a world, very near to this one, where knowledge is free, power is distributed, and agency is widespread. A world where the networked mind of humanity lifts billions from passivity into participation. A world where curiosity replaces conformity, where dignity is non-negotiable, where the game is not who wins, but how we win together.

But this world is not guaranteed, because for every force pulling toward uplift, there is a counter-force pulling toward division.

Those same tools that can empower the many can also entrench the few. When access to computation becomes the new currency, those with the most powerful models, the best data, and the closest proximity to infrastructure can lock in advantages that no longer grow, but compound geometrically.

We are already seeing the early contours of this: massive disparities in AI access, educational tools, energy storage, and biomedical enhancements. Geopolitical blocs are forming around computational supremacy. Entire populations are cut off from the tools that define modern agency, not just due to lack of funding, but due to lack of language, bandwidth, context, or inclusion. And unlike previous divides, between literate and illiterate, rich and poor, this one threatens to be existential. If the next leap in human potential shifts from social or political to biological, cognitive, or digital, then the divergence extends beyond just outcomes. It is in capacities.

➢ What happens when some can enhance memory, cognition, or creativity, and others cannot?

➢ What happens when some can interface with AI as thought partners, and others are left as passive consumers?

➢ What happens when some can live increasingly entangled with decentralized systems of knowledge and value, and others remain dependent on failing, centralized institutions?

This is the "Final Divide", not because it must end us, but because it would represent a fracture so deep that our sense of shared humanity may not survive it.

And that fracture will not be one dramatic moment. It will be subtle. Gradual. It will look like customization. Like preference. Like efficiency. The wealthiest will live in quantum-aware homes, navigate decisions with cognitive augmentation, make connections through intelligent social filters, and access medicine tailored to their genomic signatures. Others will live in the shadows of those systems, guided by algorithms they don't control, educated by fragments, employed through platforms that know more about them than they know about themselves.

And if that happens, if the great powers of this age are not designed with inclusion as a first principle, we may create not just inequality, but divergence in kind.

Two species, divided not by biology, but by access.

The question, then, is whether we will allow this. Whether we will treat these possibilities as inevitable or as invitations, because the technologies themselves do not care. They do not choose. They do not govern. We do. Design is destiny. And if we want the future to be shared, we must design it that way.

That means putting inclusion at the center of infrastructure. Building systems that are open, interoperable, and adaptable to diverse contexts. It means ensuring that the languages of tomorrow —code, data, and algorithms —are learnable, teachable, and not reserved for a priesthood of engineers.

It means thinking beyond profit. Beyond even fairness. Toward dignity, sovereignty, and participation as sacred design principles.

The Great Uplift is not a product of generosity. It is a function of architecture, and architecture can be changed.

We must ask: who is at the table when systems are built? Who decides the defaults? Who gets to train the model, define the values, write the prompt?

These are not peripheral questions. They are the essence of the future, because if we get this right, if we ensure that the new systems do not just serve the powerful, but serve the possible, we will not just avoid the Final Divide. We will achieve something far greater than uplift.

We will create a civilization in which the very definition of human potential expands, not just for some, but for all.

This is not charity. It is coherence, and coherence is the only thing that can hold complexity. We stand at the edge. The tools are in our hands. The choice is still ours.

Chapter 49: What Legacy Will We Leave?

We often think of legacy as something left behind, a monument, a record, a name etched in history. But in a quantum world, legacy is not something we leave. It's something we entangle.

Everything we build, say, touch, or design reverberates through systems larger than we can see. This is true in a poetic sense, yes, but also in a literal one. In quantum mechanics, entanglement means that two particles, once linked, can no longer be described independently. A change in one echoes in the other, regardless of distance. They remain part of the same unfolding event.

Human life is no different.

In the classical world, our legacies were linear. You taught your children, wrote your book, left your inheritance. Your influence could be measured in discrete acts. But in the quantum paradigm emerging now, legacy is a pattern, dynamic, distributed, often invisible, but no less powerful. You do not simply pass something on. You reshape the field in which others will live.

We are entangled not only with our technologies, but with time itself.

What does it mean, then, to leave a legacy in an age of planetary computation? In an era where every keystroke becomes data, every habit becomes signal, every design decision echoes across billions of lives through cascading systems we barely comprehend?

It means that intention matters more than ever.

It means that what you build, whether it's a company, an algorithm, a classroom, or a comment section, is not just a thing. It's a seed. A signal. A tuning fork. And it will vibrate long after you are gone.

This is not a metaphor. It is infrastructure. It is culture.

Legacy is being encoded into systems: smart contracts that outlive their creators, machine learning models that evolve based on the biases of their first inputs, data sets that shape future intelligence. The choices we make now, what to include, what to protect, what to silence, what to elevate, will outlive us not as history, but as active components in the future's operating system.

In this context, legacy is no longer a luxury of the powerful. It is a responsibility of the living.

And so we must ask ourselves: What kinds of ancestors are we becoming?

Because we are now ancestors not just to children or nations, but to emerging intelligences, to digital cultures, to the next iteration of civilization itself, we are shaping not only memories, but mechanisms. And the values we embed today will structure what is possible tomorrow.

This raises a deeper, more haunting question: Can a civilization become wise fast enough to survive its reach?

Wisdom is not the accumulation of information. It is the capacity to navigate complexity with care. It is the ability to see unintended consequences, to act in light of future unknowns, to prioritize coherence over conquest. Wisdom is what allows intelligence to become humane.

And legacy, in a quantum age, must be rooted in wisdom.

That means we must learn to design systems that are not only powerful, but humble. Systems that can be questioned, paused, redirected. Systems that reflect the provisional nature of our understanding. That adapts as we grow. That preserves room for voices not yet heard.

It means teaching foresight as a civic skill, designing for reversibility, embedding reflection into rapid innovation, and valuing not only invention but also the moral courage to slow down when speed becomes dangerous.

In ancient cultures, legacy was not about what you built. It was about how well you listened to the land, to the elders, to the unborn. Time was not a line. It was a loop. Responsibility did not end at death. It extended into the future as a living commitment.

Perhaps, in this quantum moment, we are being invited to remember something old by encountering something radically new.

Perhaps the most profound legacy we can leave is not a thing, but a stance. A way of seeing the world not as material to be shaped, but as a relationship to be honoured.

This is not a call to retreat. It is a call to expand. To expand the circle of concern. To broaden the scope of our imagination. To ask not just what is possible, but what is appropriate. What is just? What is beautiful? What is enduring?

Because the truth is this: the systems we are now shaping, of intelligence, economics, ecology, governance, will shape us in return. And they will shape those who come after us even more.

So let us build with memory in mind, let us code with compassion, let us teach with long echoes, let us remember that everything we touch is entangled, and that our legacies, like quantum systems, are not contained in single points, but in the fields we co-create.

We may not control the future. But we are already part of it.

The question is: what part will we play?

That is the legacy worth leaving.

Chapter 50: *The Becoming*

There comes a moment, in every long transformation, when the old maps no longer work.

You may not notice it at first. You may still follow the familiar paths, perform the rituals, speak the inherited language. But something shifts. A quiet mismatch grows between the stories you've been told and the world you're beginning to sense. The tools still function, but not quite as they used to. The rules still hold, but feel brittle. The explanations still explain, but without conviction.

This is that moment, and we are not alone in it.

Across the planet, across generations, across disciplines and traditions, there is a stirring, a collective recognition that something foundational is changing. We are not simply progressing through time, but passing through a threshold and not upgrading, but unfolding. Not just surviving but becoming.

What, exactly, are we becoming?

There is no single answer. There can't be. Because the transformation underway is not linear, and it is not singular. It is a braided river of changes, technological, ecological, existential. It is the convergence of quantum computation and planetary collapse, of machine intelligence and moral reckoning, of biological redesign and spiritual awakening.

And the name of this transformation is not yet known. But the shape of it can be felt. It is the dissolving of boundaries once thought permanent: between human and machine, self and system, thought and computation, matter and meaning. It is the recognition that we are entangled, not only with each other, but with the entire process of life, and that this entanglement is not a limitation, but a calling.

We are not stepping into a future defined by control. We are stepping into a future defined by participation.

We are no longer merely observers of complexity. We are participants in it. Shapers of it. Extensions of it, and that means we must learn new forms of agency.

In the past, the term "agency" meant power over. Power to dominate, to command, to predict, and to contain. In the coming world, agency will mean power with. Power to align, to resonate, to co-create. Power not to simplify the world into what fits our models, but to meet the world as it is, messy, relational, multi-dimensional, alive.

This shift is not only intellectual. It is spiritual. It requires a new posture of being, one that is grounded but curious. Humble, but courageous. One that can sit with uncertainty without collapsing into fear. One that can act decisively without needing control. One that can feel awe without abandoning responsibility.

Because awe is not an escape, it is an entrance. It is how we begin to relate to the scale and beauty of what is unfolding. It is how we start to ask better questions, not just of our tools, but of ourselves. Not just "What can we build?" but "Who are we becoming?" Because that is the most profound truth of all: every tool we create also creates us.

The printing press made us readers. The telescope made us stargazers. The microscope made us biologists. The internet made us networks. The algorithm is making us patterns. The quantum computer is making us wonder again.

And in this becoming, we are being invited to step into our next identity, not as masters of the world, but as conscious stewards of a reality more intricate than we ever imagined.

That is the challenge of this moment. Not to predict the future. But to prepare for it with presence. Not to control the systems. But to attune to them. Not to worship the machines. But to live alongside them, wisely. Not to abandon our humanity. But to deepen it, because what makes us human has never been our tools. It has been our stories. Our questions. Our capacity to love, to learn, to grieve, to care.

And these are not being replaced. They are being invited into a new chapter. A chapter in which civilization becomes conscious of itself. A chapter in which technology does not outpace ethics but reveals its necessity. A chapter in which the future is not something we inherit, but something we co-create, breath by breath, line by line, choice by choice.

This is not the end. This is not the beginning, either. This is the becoming. And it is ours to shape.

Together.

Epilogue

The Topological Humanity, The Space Between Worlds

We began with a wave.

A tremor felt in the distance, quiet, mathematical, and strange. Something was changing, but it didn't yet have a name. Not quantum. Not topological. Not yet civilizational. Just a sense that the world was tilting, that the future was arriving unevenly, and that if you listened closely, you could hear the early rhythms of something vast.

Now, five Sections later, we stand in the middle of the storm, not a storm of destruction, but of becoming. Ideas once confined to laboratories are reshaping institutions. Tools once reserved for experts are flowing into public hands. The line between subject and system, creator and code, self and society, has been stretched beyond recognition.

And something else has happened along the way.

We have grown.

We have learned a new vocabulary, of phase space and coherence, of open-source leverage and invisible influence. We have traced the arc from curiosity to clarity to contribution. We have dared to imagine not just what these technologies are, but what they mean. And we have remembered, again and again, that the most profound revolutions are not mechanical. They are moral. Cultural. Human.

The Shape of What's Coming has never been about technology alone. It has been about transformation. About the shape of our becoming. About the strange, beautiful, terrifying truth that we are not just adapting to the future, we are authoring it.

And now we arrive at the space between worlds.

The world we have known, based on scarcity, hierarchy, linearity, and control, is no longer sufficient. But the world we are building, distributed, dynamic, entangled, emergent, is not yet fully formed.

This space is fragile. It is precious. And it belongs to all of us.

In this liminal space, small actions echo loudly. Quiet voices shape the chorus. Experiments, stories, and communities become architectures of possibility.

Here, there is no map.

Only the question: what kind of world are we willing to create?

This is not the end of a series. It is the threshold of a movement.

A movement of conscious technologists and ethical designers. Of poetic engineers and pragmatic dreamers. Of builders, teachers, weavers, and citizens who understand that civilization is not a finished product, but a living process.

A The Shape of What's Coming is not defined by what it builds, but by how it lives, how it learns, and how it lifts.

And if you have come this far, if you have felt yourself shift, if you have seen your reflection in these pages, then you are part of that movement now.

Welcome to the work.

Welcome to the becoming.

Welcome to what comes next.

Together, we make it real.

Glossary of Key Concepts

Usage Tip

Readers may wish to revisit this glossary periodically as they move through the text, especially when encountering particularly dense or poetic passages. The glossary is not just for clarity; it's part of the conceptual toolkit the book offers for navigating transformation.

Aesthetic Ontology: Understanding the world through beauty, art, and sensory experience. It means we don't just think reality, we feel it, too.

Algorithmic Culture: A way of living where decisions, trends, and identities are shaped by algorithms, often invisibly.

Anthropic Fragility: How human-made systems can break easily, even when they seem strong, because everything depends on everything else.

Attention Ecology: The way our attention is shared, used, and shaped in a world full of distractions, like social media and ads, is fighting for our focus.

Boundary Dissolution: When the lines between things, like self and other, or human and machine, become less clear or disappear.

Civic Imagination: The shared mental space where people imagine what society could become. It helps us dream of better futures.

Code Layer: The hidden software and systems that run our digital world. It shapes how things work before we even see them.

Coherence Collapse: When a person, group, or system falls apart because its parts no longer fit or make sense together.

Collective Sense-Making: How groups of people come to understand what's happening in the world, especially in confusing or fast-changing times.

Conceptual Ruins: Old ideas or beliefs that still affect us, even though they no longer hold up or make sense.

Cultural Phase Shift: A significant change in what a culture believes, values, or builds, like going from the industrial age to the digital age.

Data Myopia: Focusing solely on numbers or data can lead to missing the bigger picture or more profound meaning.

Designing from the Future: Planning for today based on tomorrow's possibilities. Treating the future as something we shape now.

Distributed Self: A person whose identity is spread across many places, like online, at work, with friends, instead of having one clear center.

Emergent Coherence: When something organized or meaningful appears naturally from many smaller parts working together.

Epistemic Collapse occurs when people stop agreeing on basic facts, leading to a breakdown of truth.

Existential Integrity: Living in a way where your actions, thoughts, and values match, even during times of change or confusion.

First-Person Infrastructure: The inner structures that help us think, feel, and understand our lives.

Fractal Subjectivity and identity that changes depending on where you are, but still keeps its pattern, like a repeating shape.

Geometry of Belief: The way we shape and organize our beliefs is like a mental map that holds ideas together.

Ghosts of Industrial Order: Old systems and beliefs from the industrial age still influence us, even though their time has passed.

Human Topology Seeing people not as fixed things, but as flexible patterns that shift with relationships and change.

Hyperstition: A story or idea that becomes real because people believe in it and act on it.

Incoherence Collapse: When a system breaks down from the inside, it is because its parts no longer connect or work together.

Liminal Self: The version of you that exists in between stages, not who you were, not yet who you will become.

Mental Infrastructure: The invisible systems in your mind that help you make sense of the world.

Narrative Substrate: The deeper story beneath all stories is the background idea that shapes how we see the world.

Ontological Drift: The slow change in how we understand what reality is made of.

Ontological Hacking: Changing how people see reality by changing systems, language, or technology.

Paradox Literacy: The ability to hold two opposite ideas at once and still make sense of them.

Perceptual Gravity: The pull of old habits or views that keep us from seeing new possibilities.

Post-Cartesian Identity: Seeing the self as a whole being shaped by relationships, rather than being split between mind and body.

Post-Linear Time Understanding time as flexible, not just from the past to the future, but connected in loops and layers.

Pre-Real Future: A future that is starting to show up but isn't fully here yet.

Quantum Civilization: A society built on ideas from quantum science, where uncertainty, connection, and possibility are embraced.

Quantum Individual: A person who holds many roles or truths at once and can shift between them.

Reality Design: Creating environments or systems that shape how people live, think, or feel.

Semantic Drift: When the meaning of a word slowly changes over time.

Sense-Making Collapse: When our usual ways of understanding the world stop working.

Signal Integrity: Keeping a message clear and honest as it moves through noise or distraction.

Silent Revolution: A profound change that happens quietly, without protest or war, through shifts in how we think and live.

Societal Compression: The pressure builds when many significant problems co-occur in a society.

Technocultural Weather: The emotional mood is created by our constant interaction with technology and media.

Temporal Compression: The feeling that time is speeding up because of too much change or information.

Topological Crisis: A crisis in how things are connected, not just what they are made of.

Topological Intelligence: A kind of thinking that understands patterns, flows, and how things connect.

Topology of Mind: Seeing the mind as something that changes shape based on experience.

Transformation Threshold: A turning point when old ways fall apart and new possibilities begin.

Transitional Consciousness: A mindset for living through change, open to new ideas but still finding stability.

Uncomputable Dimensions: Parts of life that can't be turned into data, like love, mystery, or silence.

Vectoral Agency: The power to guide change by shaping direction, not by force.

Vertical Collapse occurs when leadership or top-down systems lose effectiveness or credibility.

Wavefront Moment: The edge of change, the moment before something big becomes visible to everyone.

References & Further Reading

The Shape of What's Coming: Why Everything Feels Broken, and What's Actually Taking Shape Beneath the Surface

This list includes foundational texts, scientific breakthroughs, philosophical works, and visionary frameworks that inform the ideas in this series. While this is not an exhaustive academic bibliography, it honours the lineage of thought that made this work possible.

- ❖ *Quantum & Complexity Science*
- ❖ *Systems, Technology & Future Design*
- ❖ *Ethics, Philosophy & Society*
- ❖ *Civic Tech, Decentralization & Web Futures*
- ❖ *Imagination, Narrative & Cultural Futures*

About The Author

Alex Nova is a systems theorist, public educator, and futurist dedicated to translating complexity into clarity. As the author of *The* Shape of What's Coming, Nova guides readers through one of the most profound transformations in civilization's history, the rise of quantum and topological technologies, and their impact on how we think, relate, and build.

With academic roots in epistemology, complex systems, and human-centred design, Mr. Nova is known for bridging the technical and the existential. Before turning to writing full-time, they advised civic labs, taught interdisciplinary design, and worked at the seams of ethics, intelligence, and infrastructure. Always a translator between worlds, Nova's passion is not just to inform, but to equip.

The Shape of What's Coming emerged from a simple but powerful conviction: that the public must not be left behind in a future being shaped without them. Across five immersive sections, Nova invites readers to grasp the tools, metaphors, and movements reshaping civilization, not with fear, but with wonder, literacy, and agency.

Mr. Nova lives between cities and silence, often off-grid and always online. He believes the most important revolutions begin in language, and that the future is not something we inherit, but something we author.

This is Mr. Nova's first major nonfiction work for a global audience.

About The Publisher

Welcome to The Book On Publishing

At The Book On Publishing, we believe in rewriting the rules of learning. Whether you're chasing your next big idea, building a better life, or simply curious about what should have been taught in school, you've come to the right place.

We're a platform built for dreamers, doers, and lifelong learners, offering bold, practical books and tools that empower you to take charge of your journey. From real-world skills to mindset mastery, we publish the book on what matters.

No fluff. No lectures. Just what you need to know, delivered with clarity, purpose, and a spark of curiosity.

Start exploring. Start growing. Start writing your story.

Read more at https://thebookon.ca.

Acknowledgment of AI Assistance

Portions of this book were developed with the support of AI. While every word has been carefully reviewed and refined by the author, AI served as a valuable tool for brainstorming, editing, and structuring ideas. Its assistance helped accelerate the creative process and bring clarity to complex topics.